Ammonia, the case of The Netherlands

Ammonia
the case of The Netherlands

edited by:

Dick A.J. Starmans

Klaas W. Van der Hoek

Wageningen Academic
P u b l i s h e r s

ISBN: 978-90-8686-028-9

First published, 2007

Wageningen Academic Publishers
The Netherlands, 2007

Table of contents

List of appendices

Glossary

	English	Dutch
AMvB	General order in counsil	Algemene maatregel van bestuur
ASG	Animal sciences group of WUR	
BIN	Dutch farm accountancy data network	Bedrijven informatie netwerk
BREF-ILF	Best available techniques reference document on intensive rearing of poultry and pigs	Referentiedocument over de best beschikbare technieken voor het intensief houden van kippen en varkens
CBS	Statistics Netherlands	Centraal bureau voor de statistiek
CEC	Cation exchange capacity	Uitwisselingscapaciteit voor kationen
DOAS	Differential optical absorption spectroscopy	Differentiële optische absorptie spectroscopie
EC	Electrical conductivity	Geleidbaarheid
EEA	European environment agency	Europees milieu bureau
EF	Emission factor	Emissiefactor
EU	European Union	Europese Unie
FOMA	Financial board on manure and ammonia research	Financieringsorgaan mest en ammoniak
FTIR	Fourier transformed infrared spectroscopy	Fourier getransformeerde infrarood spectroscopie
GAMS	General algebraic modelling system	Algemeen algebraïsch modelleersysteem
HTU	Height of transfer unit	Hoogte theoretische overdrachtseenheid
IPPC	Integrated prevention and pollution control	Geïntegreerde preventie en controle op vervuiling
LBT	Annual agriculture census	Landbouwtelling
LEI	Agriculture economics research institute of WUR	Landbouw economisch instituut van WUR
LNV	Ministry of agriculture, nature and food quality	Ministerie van landbouw, natuur en voedselkwaliteit
LRTAP	Long range transboundary air pollution	Grensoverschrijdende luchtvervuiling over lange afstand
MAM	Mineral and ammonia model	Mineralen en ammoniak model
MAMBO	Mineral and ammonia model for policy support	Mineralen en ammoniak model voor beleidsondersteuning
MEGISTA	Working group on animal manure, liquid manure and odour	Werkgroep mest, gier en stank
MINAS	Mineral accounting system	Mineralen administratie systeem
MNP	The Netherlands environmental assessment agency	Milieu en natuur planbureau
NEC	National emission ceiling	Nationaal emissieplafond

NTU	Number of transfer units	Aantal theoretische overdrachtseenheden
OECD	Organisation for economic cooperation and development	Organisatie voor economische samenwerking en ontwikkeling
PFS	Passive flux sampler	Passieve flux sensor
RAV	Guideline for ammonia and animal husbandry	Richtlijn voor ammoniak en veehouderij
Rav	Ammonia and Livestock Farming Regulation	Regeling voor ammoniak en veehouderij
RIVM	National institute for public health and the environment	Rijksinstituut voor volksgezondheid en milieu
RIZA	Institute for inland water management and waste water treatment	Rijksinstituut voor integraal zoetwaterbeheer en afvalwaterbehandeling
SDL	Regulation stimulating sustainable agriculture	Regeling stimulans duurzame landbouw
TAC Rav	Technical advisory committee for the Rav	Technische adviescommissie Rav
TAN	Total ammoniacal nitrogen	Totale hoeveelheid ammoniak stikstof
TDL	Tunable diode laser	Tunable diodelaser
UAV	Regulation for implementation ammonia and animal husbandry	Uitvoeringsrichtlijn ammoniak en veehouderij
UNECE	United Nations economic commission for Europe	Economische commissie voor Europa binnen de Verenigde Naties
UUC	Urinary urea concentration	Ureum concentratie in urine
V&W	Ministry of transport, public works and watermanagement	Ministerie van verkeer en waterstaat
VOC	Volatile organic compounds	Vluchtige organische verbindingen
VROM	Ministry of housing, spatial planning and the environment	Ministerie van volkshuisvesting, ruimtelijke ordening en milieubeheer
WB	Willems badge	Willems badge
WEM	Working group emission monitoring	Werkgroep emissie monitoring
WUM	Working group uniform mineral and manure excretions	Werkgroep uniformering berekeningswijze mest- en mineralencijfers
WUR	Wageningen university and research centre	Wageningen universiteit en research centrum

Preface

In approximately 200 pages, this book presents a condensed overview of more than 25 years of Dutch research on ammonia emissions associated with animal production. For more than a century there has been an interest in nitrogen losses from storage of farm yard manure and from manure application. This interest was triggered by early practical experience and scientific evidence that nitrogen was a valuable plant nutrient.

Around 1980 it became clear that emissions of ammonia, sulphur dioxide and nitrogen oxides played a major role in the formation of acid rain, which leads to acidification and eutrophication of soils and surface waters. From that moment on, both national and international research focussed on all aspects of acid rain. Starting with an inventory of the emission sources of sulphur dioxide, nitrogen oxides and ammonia, their respective source strengths, atmospheric transport and subsequent deposition behaviours were studied, as well as their effects on sensitive nature preservation areas and dose - effect relationships.

This book however, will only deal with a single aspect: the ammonia emissions arising from animal production. This simplification is by no means random. Before ammonia became apparent as an environmental issue, it was clear that the total amount of animal manure in The Netherlands posed an equally important environmental threat. It has lead to the close connection between scientific research and governmental policy on animal manure and ammonia emissions.

This book will give an overview of the research and policy efforts to tackle the ammonia emissions in The Netherlands. Chapter 1 provides a brief overview of animal production in The Netherlands and Dutch legislation directed to animal production, whereas Chapter 2 describes the national and international policy with respect to ammonia emissions. Chapter 3 provides a theoretical overview of emission processes and Chapter 4 describes the current model for calculation of the ammonia emissions in the framework of the Dutch National Emission Inventory. Emission abatement techniques for practical cattle, pig and poultry houses are described in full detail in Chapter 5, which also holds data on abatement techniques related to animal manure application. Chapter 6 is focused on techniques and protocols to measure ammonia emissions. Finally, Chapter 7 gives an overview of transitions in animal production and concludes with a forecast for 2030.

An impression of current and low emission animal housing systems in The Netherlands is given in the form of fact sheets for a number of representative housing systems. These are included as appendices to this book.

This book has been made financially possible by the Dutch Ministry of Agriculture, Nature and Food Quality via their Research program Gaseous emissions from livestock farming systems.

All authors of the chapters are actively involved in ammonia research in animal production settings. It is clear that many more contributed to ammonia science; their works and names are kindly and carefully referred to. We all hope that this book will give you a good overview of all matters concerning Ammonia, the case of The Netherlands.

Dick A.J. Starmans and Klaas W. Van der Hoek

1. Animal production and related environmental aspects in The Netherlands

Dick A.J. Starmans and Klaas W. Van der Hoek

The Netherlands are known for their high animal density and high animal production. After a brief description of The Netherlands, an overview of the specific Dutch situation regarding animal production and fertiliser use is given and compared to other EU Member States. The second part of the chapter discusses briefly the efforts of both research and policy regarding animal manure and ammonia emissions.

1.1. Introduction to The Netherlands

The Netherlands are a flat country, with some hilly regions in its Eastern and Southern parts. The highest point (Vaalserberg) is found at 321 m above sea level, whereas the lowest is at 6.74 m below sea level (Nieuwerkerk aan de IJssel). About half of The Netherlands would be flooded by the sea without the protection of dunes and dikes.

Due to its location between 51 and 54° North latitude and its proximity to the North Sea, the climate in The Netherlands is mild, characterised by cool summers and soft winters. The annual precipitation exceeds 800 mm and is distributed equally over the entire year. The temperature averages a mild 10 °C, with extremities in January (-1.7 °C) and July (21 °C).

Soil conditions in The Netherlands are divers and range from light sandy soils in the higher regions in the Eastern, middle and Southern parts, to heavy clay in the North and Western parts. Deposition of light clay along the major rivers explains the presence of arable and fruit farming activities in these areas. Peat is found in several places in the West and North, which makes good soil for the cultivation of grass and vegetables. Along the coast, the soils are sandy and deprived of lime. Due to local hydrological circumstances with water coming from the dunes, these soils could be exploited for local horticulture. Levelled dunes proved to be a good location for the development of bulb cultivation.

The mild climate, the reasonably fertile soil conditions and the generally steady supply of water have lead to the build-up of a healthy agriculture with a large diversity of products. By facilitating the import of feed rations, main port Rotterdam has stimulated animal production to become of great importance for the agriculture sector in The Netherlands.

1.2. Animal production

Animal production has drastically increased in Western Europe during the past decades, particularly in The Netherlands. The number of farm animals rapidly increased especially after the 2nd World War, due to mechanisation, specialisation and EU price support. For example, the annual growth rate of the number of pigs in the late sixties was about 10%. By that time, animal farming became one of the major factors contributing to the gross net income of The Netherlands. As shown in Figure 1.1, high animal densities were reached in The Netherlands from the eighties of the past century onward. Dutch agriculture has intensified thanks to large inputs of relatively cheap chemical fertilisers and concentrates.

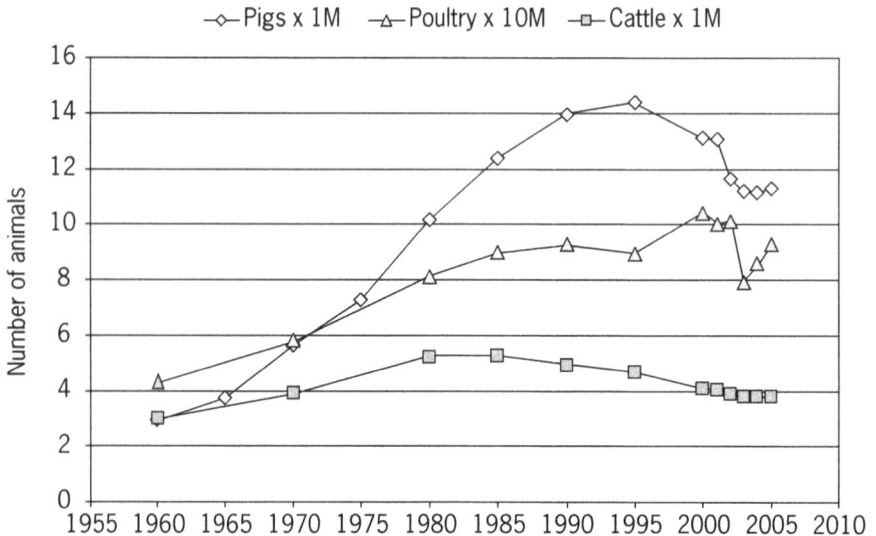

Figure 1.1. Livestock population in The Netherlands (CBS-Statline, 2006).

The apparent 10% decrease in the number of cattle from 1980 to 1995 as given in Figure 1.1 is caused by the introduction of dairy milk quota and low meat prices due to EC regulations on subsidies on extensive meat production. The decrease in pig numbers in 1997 is the result of the outbreak of classical swine fever in that year. The decrease in poultry numbers in 2003 is due to the avian flue in that year.

Most of the animal manure in The Netherlands is in liquid form as can be seen in Table 1.1. Dairy cattle is kept in animal houses where liquid manure is stored mainly underneath the slatted floors. To get through the winter period, farmers typically build an outside storage facility for liquid manure. During summertime the cattle spends some time in the meadow. Beef cattle and veal calves are also kept in animal houses with liquid manure systems, whereas suckling cows are mostly kept in houses with solid manure systems. Other ruminants such as sheep, milking goats, horses and ponies are predominantly kept in animal houses with solid manure systems.

Pigs and poultry are held in animal housing all year round. Nearly 100% of the pigs are held in animal houses with liquid manure systems, with manure stored in the animal house itself and in outside storage facilities. With the exception of laying hens, poultry produces only solid manure. During the last 25 years, farms with laying hens (both under 18 weeks, and 18 weeks and over) have switched from almost 100% liquid manure to over 90% solid manure systems.

The Netherlands with a total agricultural area of 2,000,000 hectare are characterised by a high animal density and a high use of fertiliser. Amidst other European countries it ranks highest with respect to nitrogen loading per hectare of agricultural land. Table 1.2 presents the figures for all EU15 member states for the year 2000. The high use of fertiliser on grassland is one of the reasons that The Netherlands use about 2.4 times more fertiliser nitrogen per hectare than the average value of 73 kg per hectare for the EU15 member states. The high animal density in The Netherlands is responsible for a four times higher manure nitrogen production per hectare than the EU15 average value of 66 kg per hectare of agricultural area.

Table 1.1. Animal manure production in The Netherlands, classified into main animal categories and type of manure management, with production figures given in 10^9 kg (Van der Hoek and Van Schijndel, 2006).

Animal category	Type of manure	1990	1995	2000	2003
Dairy cattle	total manure production	57.83	52.75	49.06	47.08
	liquid stable manure	38.64	35.24	33.84	34.80
	solid stable manure				
	meadow	19.19	17.51	15.22	12.28
Beef cattle and veal calves	total manure production	8.40	8.92	7.68	6.77
	liquid stable manure	5.98	6.01	4.83	4.22
	solid stable manure	0.84	1.02	1.14	1.01
	meadow	1.58	1.89	1.71	1.54
Other ruminants	total manure production	2.47	2.70	2.72	2.66
	liquid stable manure				
	solid stable manure	0.65	0.81	0.94	1.03
	meadow	1.82	1.89	1.78	1.63
Pigs	total manure production	16.36	16.15	14.13	11.72
	liquid stable manure	16.36	16.15	14.13	11.72
	solid stable manure				
	meadow				
Poultry	total manure production	2.59	2.17	2.19	1.23
	liquid stable manure	1.45	0.90	0.53	0.16
	solid stable manure	1.14	1.27	1.66	1.07
	meadow				
Total liquid stable manure		62.4	58.3	53.3	50.9
Total solid stable manure		2.6	3.1	3.7	3.1
Total manure in meadow		22.6	21.3	18.7	15.4
Total manure production		87.7	82.7	75.8	69.5

Ammonia, the case of The Netherlands

Table 1.2. Fertiliser use and animal manure production in the EU15 member states in 2000 (Eurostat: agricultural areas; EAA, 2006: fertiliser and manure data).

	Agricultural area 10^3 hectare	Nitrogen fertiliser 10^6 kg N	Animal manure 10^6 kg N	Nitrogen fertiliser kg N/ha	Animal manure kg N/ha	Nitrogen fert + animal manure kg N/ha
Austria	3,407	117	164	34	48	83
Belgium	1,396	157	327	112	234	346
Denmark	2,641	246	270	93	102	195
Finland	2,209	166	112	75	51	126
France	29,796	2,314	1,961	78	66	143
Germany	17,067	2,014	1,346	118	79	197
Greece	3,901	257	408	66	105	170
Ireland	4,418	399	457	90	103	194
Italy	15,189	717	935	47	62	109
Luxemburg	135			0	0	0
The Netherlands	1,969	339	528	172	268	440
Portugal	3,907	140	165	36	42	78
Spain	25,386	1,264	837	50	33	83
Sweden	2,974	189	159	64	54	117
UK	15,722	1,202	919	76	58	135
EU15 states	130,117	9,522	8,590	73	66	139

1.3. Supporting role of research

From 1985 onward, the Ministry of agriculture, nature and food quality (LNV) took the initiative for a large research program focusing on solving the environmental problems associated with the intensive animal husbandry in The Netherlands. Funds were raised to finance the research by a joint venture between the Ministry of housing, spatial planning and environment (VROM) and the farmers' agricultural associations (called 'Landbouwschap' and 'Productschappen').

Between 1985 and 1995, over 200 million Dutch guilders (90 million Euros) were invested in research projects in the fields of:
• environmental assessment;
• animal nutrition;
• manure utilisation on grassland and arable land;
• manure handling on farm scale;
• central processing of manure;
• manure transport and logistics;
• economic evaluation.

This research, co-ordinated by the Financial board on manure and ammonia research (FOMA), resulted in a 'knowledge boost' in the field of agriculture and environment, and formed the basis of most of the legislation we have nowadays.

The research and development also facilitated farmers and industries to meet environmental goals by using innovative technical solutions generated. In the field of nutrient and ammonia legislation, the FOMA-program laid the foundation for a number of subsequent research programs initiated and stimulated by the Ministry of LNV. This however, did not mean that FOMA actually solved the entire problem of nutrient and ammonia losses to the environment. Although it did yield many solutions or strategies in the fields describe above, the most important result was the facilitation of a transition in agriculture from increasing intensification (and pollution) towards a more balanced and sustainable development in general, and animal husbandry in particular. The role of research will be discussed in more detail in Chapter 2.

1.4. Environmental legislation in relation to animal production

The enduring increase of intensive livestock production (pigs, poultry) especially in the South and South-East (sandy soils, the so-called concentration areas) of The Netherlands together with signs from the research community (see Chapter 2) lead to subsequent environmental legislation which started in November 1984 with the Interim Law pigs and poultry.

1984: Interim law pigs and poultry
The goal of this law was restricting pig and poultry farming in threefold:

- No new farms.
- Maximal expansion of 10% on existing farms in concentration areas.
- Maximal expansion of 75% outside these areas.

The interim law was originally intended to be active for the limited period of two years. It was intended to pave the way for two framework-laws that would become the foundation for further legislation (through the issuing of orders in counsel (AMvB)). These laws arrived in 1986 and 1987.

1986: Use of Fertilisers Decree ('Meststoffenwet')
The acknowledgement of the Dutch government that the amount of animal manure produced exceeded the amount of animal manure that could be applied on agricultural land in a sustainable way, made good regulation of animal manure matters necessary. The following points were taken up in this law:
- Control of animal manure production by regulation of the number of animals: expansion and starting of new livestock farms is prohibited and annual animal countings are performed.
- National manure bank to enhance the efficiency of transport of animal manure to arable farms.
- Control of transports by government officials.

1987: Use of Animal Manure Decree (by virtue of the Soil Protection Act) ('Wet bodembescherming')
This law was put in order to protect all soils from all kinds of pollution. With regard to animal manure, this law regulated the amounts of manure application onto soils, the periods when these amounts were allowed to be applied and finally which methods of application were approved.

These both laws heavily depended on research to enhance the technological possibilities of maintenance and control. The responsibility to do so was taken up by the ministry of LNV. Moreover, the already existing Public Nuisance Act ('Hinderwet') was expanded into a system with permits for animal production that had to include local nature and environment as well.

A report by the Institute for research on plant diseases increased the awareness of direct damage of ammonia emitted from animal husbandry on trees and other types of vegetation (Van der Eerden *et al.*, 1981). At the same time it became apparent that the environmental problems related to the emission and deposition of ammonia played a major factor in the acid rain issue. From that moment on, ammonia emissions were put firmly on the political agenda. The following legislation on ammonia emissions was issued in subsequent years.

1987: Guideline for Ammonia and Animal Husbandry ('Richtlijn Ammoniak en Veehouderij')

This guideline was issued as part of the revised Public Nuisance Act. It was accompanied by Order in council ('AMvB') on manure storages, which stated that outside manure storages near acidification sensitive areas had to be covered. The guideline was preliminary taken up in legislation in 1994.

1994: Ammonia and Livestock Farming (Interim Measures) Act ('Interimwet Ammoniak en Veehouderij')

The interim law included a Regulation for Implementation ('Uitvoerings-richtlijn Ammoniak en Veehouderij. UAV'). As part of this law, a vast amount of money was invested in research and development of low emission techniques in general, and low emission housing systems in particular. Full legal status of this law was achieved in 2002.

2002: Ammonia and Livestock Farming Act ('Wet Ammoniak en Veehouderij')

The guideline was replaced by the Ammonia and Livestock Farming Regulation ('Regeling voor Ammoniak en Veehouderij, RAV').

Since 1993, farmers and industry were financially encouraged to develop and implement low emission housing systems on a voluntary basis. Systems with a 50% reduction (as compared to traditional housing) received a Green Label award and were taken up in the regulation. As a benefit to those embracing these environmentally friendly systems the government allowed a shorter write-off period of Green Label housing systems. As a consequence, more than 50 Green Label Awards were issued and the implementation of low emission housing systems was accelerated, especially in pig and poultry husbandry. This development was in contrast to dairy farms, where few technical solutions were developed and implemented. Instead, nutritional measures were brought forward in this sector, with milk urea content as an important indicator.

1.5. Ammonia, the case of The Netherlands

This chapter presented a brief overview of animal production and associated environmental legislation in The Netherlands. To cover *every* aspect of ammonia, would lead the reader along many topics ranging from emissions from synthetic fertilisers, chemical industry and other non-agricultural emission sources to atmospheric processes and deposition and effects on sensitive nature preservation areas, critical loads and monitoring of air quality and quality of these areas. This was however not the goal of this book, nor was the book intended to provide evidence of similarity between calculated and measured emission patterns.

This book is devoted to ammonia emissions that arise from animal production. In subsequent chapters we will present national and international ammonia policy, the Dutch national ammonia emission inventory, low emission solutions for animal housing and animal manure application and how to measure these. The book concludes with an outlook towards animal production in The Netherlands in 2030.

References

CBS-Statline, 2006. Data available via internet: http://statline.cbs.nl.

EEA, 2006. Annual European Community greenhouse gas inventory 1990 - 2004 and inventory report 2006. EEA Technical Report 6/2006. EEA, Copenhagen, Denmark.

Van der Eerden, L.J.M., H. Harssema and J.V. Klarenbeek, 1981. Stallucht en planten. IPO Report R-245. Instituut voor plantenziektenkundig onderzoek, Wageningen, The Netherlands.

Van der Hoek, K.W. and M.W. Van Schijndel, 2006. Methane and nitrous oxide emissions from animal manure management, 1990 – 2003. Background document on the calculation method for the Dutch national inventory report. RIVM Report 680125002; MNP report 500080002. RIVM / MNP, Bilthoven, The Netherlands.

2. National and international ammonia policy

Klaas W. Van der Hoek

Research and development form the scientific corner stones in formulating and implementing new policy and legislation. The contribution of science can be divided in an exploring and a supporting role. The exploring role is focussed on finding solutions to policy related problems. The first step is characterisation of the problem, followed closely by exploration of promising solutions and options to minimise or even prevent the problem from occurring. The supporting role is focussed on assisting in formulating the necessary legislation and associated targets and on monitoring the effectiveness of policy measures. The role of science gradually expands from purely exploratory and abstract into more supporting and involved. This chapter describes both roles of science with respect to policy on ammonia emissions. Nowadays, science is also supporting international negotiations.

2.1. The exploring role of science

Ammonia is an animal manure related issue and its role as an environmental threat became apparent in the early 80's. Already in the middle of the 60's there were signs indicating that animal manure might have negative impacts on the environment. Because of the close relation between animal manure problems and ammonia emissions, this section gives an overview of the role that science has played in recognising and accepting the animal manure problem in The Netherlands. Science also had an exploring role in finding technical solutions to deal with the animal manure problem, or at least to indicate which pathways seemed promising for solving this problem.

In view of the increasing production of animal manure, experiments with aerobic purification started on farm scale as early as 1965. The project was subsidised by the Ministry of transport, public works and water management. The results showed that highly diluted manure, such as veal calf manure with 1-2% dry matter, could be treated. However, the resulting effluent appeared not to be suitable for sewer discharge. Experiments with liquid pig manure failed due to the high dry matter content, which is normally around 8%. The amount of effluent turned out to be very small and it contained too much oxygen demanding components (Ten Have, 1971). During the pig manure treatments it was observed that the aerobically treated manure had a low level of odorous components, which is an advantage during manure spreading. In later years it was also observed that anaerobic treated animal manure had a low level of odorous components, especially because of a lowered concentration of volatile fatty acids.

In the same period farm scale experiments started with biological air scrubbers. The objective was to remove the odorous components from the exhaust ventilation air from animal houses. Good functioning biological air scrubbers were a prerequisite to remove odour successfully. Although biological air scrubbers were also very capable to remove ammonia, the focus was not on ammonia (Van Geelen and Van der Hoek, 1982).

Also around 1970 the scientific discussion started about the criteria to establish the maximum amount of animal manure per hectare. For grassland the component potash was chosen as limiting element, because too much potash appeared to be harmful for cattle. For arable production nitrogen was considered as the limiting element. Too high nitrogen application rates were found to be harmful for cereals and potatoes (Henkens, 1975). On the other hand renewed interest for the fertiliser value of animal manure was directed at the nitrogen components. Part of the nitrogen is in organic form and is slowly released over the years after manure application. The other part (ammonium) is in inorganic form and is readily converted into nitrate and available for the growing crop. Knowledge of the time-based plant availability of both organic and inorganic nitrogen allows a better control of the optimal amount of nitrogen fertiliser needed for grass and crop production (Sluijsmans and Kolenbrander, 1977).

An early research into the agronomic value of animal manure was summarised and discussed (Kolenbrander and De La Lande Cremer, 1967). Their extensive review discussed the amounts of manure produced per animal, the mineral composition, the storage of farm yard manure and liquid manure, and the fertiliser value of animal manure. They also paid attention to nitrogen losses during storage and application of the manure. Nitrogen losses during land spreading depended on the time span between application and incorporation by ploughing. Figure 2.1 shows the nitrogen loss, presented as a decreasing nitrogen use efficiency. The efficiency is about 50% when the applied farm yard manure is incorporated immediately, thus concluding that a part of the nitrogen is in organic form and not readily available for the crop production.

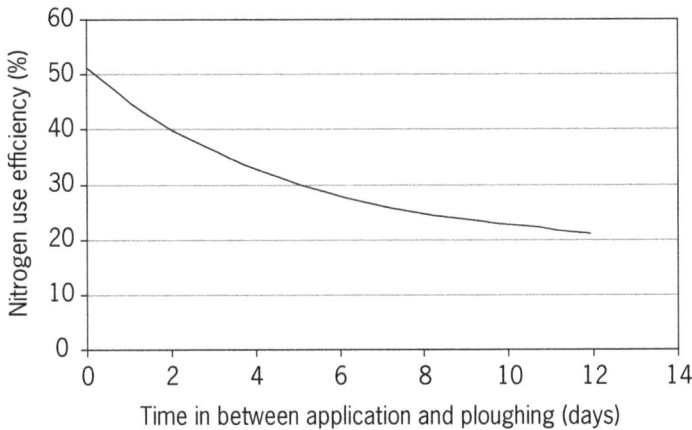

Figure 2.1. Effect of time in days between application and ploughing of farm yard manure on the nitrogen use efficiency (Kolenbrander and De La Lande Cremer, 1967).

Animal manure application in excess of crop nitrogen uptake may lead to nitrate leaching towards lower water bodies. It takes quite some time before nitrate enters the water layers used for extracting public drinking water. The first signs of excess manure application were found in the pumping stations of Montferland and Reuver, both located in the sandy areas of The Netherlands (Figure 2.2; Van Beek *et al.*, 1984). Model calculations showed that the nitrate concentrations would steadily increase after 1985 because of the continuation of the excess manure application and the time-scale at which the nitrate leaching process would progress.

A report of the Institute for research on plant diseases demonstrated the direct damage of ammonia on trees and other types of vegetation (Van der Eerden *et al.*, 1981). Ammonia as an environmental threat became evident in 1982 when high levels of ammonium sulphate were found in rainwater. The high levels of ammonium sulphate resulted in increased soil acidification because ammonium sulphate was rapidly oxidised into nitric and sulphuric acid. Normally, the uptake of nitrate ions by growing plants is accompanied by the uptake of H^+ ions. Soil acidification takes place when surplus ammonia reacts to nitrate and H^+ ions (Van Breemen *et al.*, 1982).

The research community regarding animal manure related problems was assembled in November 1972 in MEGISTA, a Dutch acronym for 'mest,

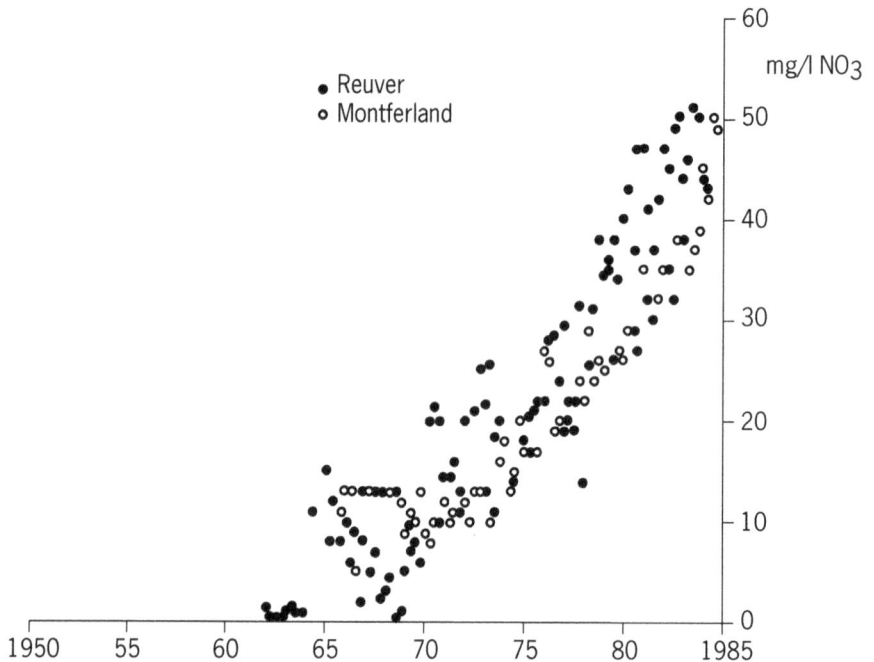

Figure 2.2. Increase of nitrate content of extracted drinking water for two locations (Van Beek et al., 1984).

gier en stank' (animal manure, liquid manure and odour). The first general assembly took place in the spring of 1973, where ongoing research was presented and further research priorities for the coming period were formulated (Huisman, 1973).

2.2. Dutch ammonia policy

The Dutch legislation on ammonia and animal manure is closely interrelated. The bird's-eye view of the legislation described in Section 1.4 resulted in the following current ammonia policy of the Dutch government:

- *Animal housing.* Following the Ammonia and Livestock Farming Act (2002), every farm needs a permit describing the maximum ammonia emissions allowed. This defines the farm size, because of the known emission factors for all traditional and low emission animal houses. The legal obligation to move to only low emission housing systems for pig and poultry is foreseen for 2010 but not yet in legal force. The maximum allowed emission factors for pig and poultry houses are presented in Table 2.1.

- *Animal feed.* Following the Ammonia and Livestock Farming Act, 2002, an agreement was made with LTO (Dutch Farmers Union) that a low milk urea content in dairy farming could replace the building of low ammonia housing systems for dairy cattle (see Section 5.1.3 feeding management).
- *Outside storage of animal manure.* Following the Public Nuisance Act, covering of all outside storage facilities for liquid manure became obligatory in 1987. Natural crust was not allowed as a legal cover (see Section 2.3.3).
- *Land spreading of animal manure.* Following the Use of Animal Manure Decree (by virtue of the Soil Protection Act) in 1987, surface spreading of animal manure was no longer allowed in the first part of the growing season on sandy areas. Gradually surface spreading of animal manure was banned in the whole of The Netherlands and in 1995 nearly all manure was applied with low emission techniques.

Table 2.1. Maximum allowed emission factors for pig and poultry houses, as foreseen for 2010, but not yet in legal force. For comparison emission factors for current traditional housing systems are presented.

	Emission factor kg NH_3 (animal place)$^{-1}$yr^{-1}	
Animal category[1]	Maximum	Current
Weaned piglets < 25 kg	0.23	0.60 – 0.75
Farrowing sows (incl. piglets)	2.9	8.3
Dry and pregnant sows	2.6	4.2
Fattening pigs	1.4	3.0 – 4.0
Layers < 18 weeks held in cages	0.006	0.020 – 0.045
Layers and layer breeders, held in cages	0.013	0.035 – 0.083
Layers and layer breeders, held in non cages	0.125	0.315
Broiler breeders	0.435	0.580
Broilers	0.045	0.080

[1]See tables 5.3 – 5.6 for more information on emission factors for animal housing systems for pigs and poultry.

• *Restricted area of farming close to nature preservation areas.* Following the Ammonia and Livestock Farming Act, 2002, new farms are no longer allowed in a zone of 250 meter around nature preservation areas vulnerable for ammonia. Existing farms in this zone may only expand if housing systems with very low ammonia emission rates are used.

2.3. The supporting role of science

This section describes the scientific support in three domains: emissions from animal houses, emissions from the other ammonia sources like animal manure application and finally the translation from research into national emission inventories.

2.3.1. Emissions from animal houses

For a successful manure and ammonia policy reliable data on manure production and its mineral composition were a prerequisite. Around 1986 the Dutch government had chosen for phosphate as the element for estimation of the manure production per animal on the one side and the maximum amount of manure per hectare on the other site[1]. Therefore for all animals counted in the annual agricultural census calculations were made of the annual phosphate (P_2O_5) excretion. The following equation was used:

$$P_2O_5 \text{ excretion} = 2.29 \times (amount\ of\ feed \times P\ conc - amount\ of\ products \times P\ conc) \quad (1)$$

2.29 is a factor to convert P into P_2O_5, all amounts are in kg per year

The amounts of feed and products like meat, milk and wool were obtained from various sources: statistics, handbooks, and the Dutch farm accountancy data network (Van der Hoek, 1987). Following this approach the nitrogen excretion was calculated. By comparing the N/P_2O_5 ratio in the excreted manure and the N/P_2O_5 ratio in stored animal manure the loss of nitrogen during storage of the animal manure was estimated (De Winkel, 1988). This approach is presented in Equation 2.

---

[1] Phosphate (P_2O_5) was chosen as key element because it does not escape from the agricultural environment. Nitrogen in contrast can easily escape as ammonia, nitrous oxide, dinitrogen thus leading to difficulties in setting allowable amounts of nitrogen. In later years nitrogen was also included as key element in the legislation. The N over P_2O_5 ratio in animal manure varies between animal categories and therefore additional nitrogen standards were necessary as nowadays included in the EU Nitrate Directive.

$$NH_3 \text{ emission} = 17/14 \times (P_2O_5 \text{ in manure} \times N \text{ excr}/P_2O_5 \text{ excr} - N \text{ in manure}) \qquad (2)$$

The formula gives in fact the NH_3-N emission, so the factor 17/14 is used for conversion to NH_3

The ammonia emission values were included in the Guidelines for Ammonia and animal husbandry, to be used for official permits for animal production. In the accompanying explanatory appendix it was noted that these values were estimated values to be replaced in due time with measured values.

The ammonia emission measurements were done on farm scale and were subjected to described procedures. A working group made an inventory of available measurement techniques suitable for use under farm conditions (Van Ouwerkerk, 1993). This aspect is elaborated in detail in Chapter 6.

In order to obtain comparable results, the conditions during the measurements had to be standardised. The first step in this standardisation was to find a suitable length of the measurement period in order to obtain results representative for the emissions during the whole year. Secondly, the agronomic conditions inside the animal house (such as temperature) had to be representative for the animal category. In the third place, also the zootechnical conditions had to be representative (like protein content of the feed) for the animal category. A protocol for measurements (Beoordelingsrichtlijn) was delivered by another working group (Anonymous, 1993). The dates of publication are somewhat misleading because most of the content of both documents was available before. The majority of measurements were carried out by the so-called Housing measurement team (Stalmeetploeg). Some measurements were done by private firms.

2.3.2. Support in the evaluation of new housing systems

Around 1991 it was felt that the development of new housing systems with low emission rates could be encouraged by granting Green Label awards (Figure 2.3). Manufacturers of animal housing systems could safeguard their investment costs for a new product because the Green Label award was an advantage on the market. For the individual farmer it was an advantage to buy a Green Label housing system because of the fiscal profits in terms of extended depreciation possibilities and the farmer got the promise not to be forced to invest in newer emission reduction techniques in the coming years.

Figure 2.3. Logo of Green Label award for low emission animal housing systems. Green Label was succesful in generating and introducing new animal housing systems in The Netherlands.

The Green Label awarding became operational in January 1993 with a Board, a Technical Advisory Committee and a secretariat. Applications for a Green Label award were reviewed by the Technical Advisory Committee and the award was granted by the Board. The review procedure was based on the protocol for measurements (Beoordelingsrichtlijn). For every award a leaflet was made with the description of the animal housing system, the unique principle leading to the emission reduction, the obtained emission factor and checkpoints for verification of good functioning. In 2001 the Green Label awarding system was terminated and plans were made for a new certification system with a broader scope including energy and animal welfare (SDL, Regeling Stimulans Duurzame Landbouw).

The working group Emission Factors was responsible for the review of the ammonia emission measurements. Not only the report of the measurement was studied, also the basic principle for emission reduction and a clear definition of the animal housing system were discussed. Finally the working group prepared an advice to the Ministry of Environment for uptake of a new emission factor in the Regulation for implementation ammonia and animal husbandry (Uitvoeringsrichtlijn ammoniak en veehouderij).

2.3.3. Emissions from other ammonia sources

In The Netherlands animal manure is primarily stored in the animal house. Due to the ban on land application of animal manure outside the growing season, farmers extended their storage capacity with outside storage facilities. Already in the years around 1987 it was obligatory to cover the outside storages for liquid manure. The emission reduction of different cover materials was investigated (De Bode, 1991). The effect of a natural crust on the liquid manure was also tested but it turned out to be insufficient and unreliable in comparison with fixed roofs. More specific it is not guaranteed that under all weather conditions the formation of a natural crust is guaranteed.

Similar to the emission measurements from animal housing, a special Field measurement team (Veldmeetploeg) was formed to measure ammonia emissions during application of liquid animal manure. Manure was applied using different low emission techniques on a small circular area and the ammonia emission was measured with two masts, one in the middle of the field and the other outside the field (see Section 5.4.4).

A part of the nitrogen in manure deposited during grazing in the meadow is volatilised as ammonia. The emission rate depends on the amount of nitrogen excreted by the animal, the weather conditions and the soil characteristics. The amount of nitrogen excretion is dependent on the nitrogen content of the grass and the time the animals spend in the meadow (see Section 5.1.1) (Bussink, 1996).

2.3.4. Translation of research into national emission inventories

Emission inventories are usually based on a simple formula by multiplying an activity with an emission factor and summing up all activities. This means that all emission sources have to be known, the magnitude of every source (= activity) must be known and last but not least a representative emission factor is necessary.

The construction of emission factors for the different activities requires insight in the underlying animal categories and animal housing systems. Measurements are valid for the conditions during the measuring period and for local circumstances. Knowledge how these conditions influence the emissions is necessary for up scaling to national level. Discussion and interaction between researchers and emission inventory experts is the basis for a sound and reliable emission inventory (Van der Hoek, 1994, 2002a,b).

For ammonia emission inventories an activity is for instance animal housing of cattle. The emission is the number of cattle times an emission factor for cattle houses. The number of cattle is available from the annual agricultural census. The emission factor for cattle houses is the average value of all types of housing systems, taking into account the different levels of milk production and the time animals spend in the meadow. The Netherlands use a nitrogen flow model for calculation of the ammonia emissions. Nitrogen is excreted in the animal house, ammonia emissions that occur during housing and storage are subtracted and the remaining nitrogen flow is intended for land spreading. Emissions are expressed as a loss percentage of the nitrogen flow. The advantage of this nitrogen flow model is that the results of measures in one compartment become directly visible in another compartment. The Dutch ammonia emission model is discussed into more detail in Chapter 4.

2.4. International ammonia policy

Atmospheric transport of gaseous emissions crosses country borders and emissions originating in one country can be deposited in another country. In general individual countries are not able to reach deposition targets by national legislation alone. In 1979 the Member States of the United Nations Economic Commission for Europe, shortly UNECE, adopted the Convention on Long-range Transboundary Air Pollution (LRTAP). On 30 November 1999 the Protocol to abate acidification, eutrophication and ground-level ozone was adopted in Gothenburg (Sweden). At the end of 2006 the Gothenburg protocol was signed by 31 parties and ratified by 21 parties, including The Netherlands. More specific information on the protocol is presented in the Box 2.1.
The 2010 emission ceiling for The Netherlands is set at 128 million kg ammonia. This figure comprises the ammonia emissions from all sectors. Agriculture is responsible for about 90% of all ammonia emissions.

Within the framework of the Convention on LRTAP, two expert groups are dealing with ammonia:
• The agriculture and nature expert panel residing under the Task Force Emission Inventories and Projections, contributes to the corresponding chapters in the Atmospheric Emission Inventory Guidebook, issued by the European Environment Agency in Copenhagen (Anonymous, 2005). This Guidebook is intended to support countries with compiling of their national emission inventories. Dutch scientists play an active role in this expert panel (Van der Hoek, 1998).
• The expert group on ammonia abatement residing under the Working group on strategies and review is responsible for feasible options for ammonia emission reduction. Their work results in a Code of Good agricultural practice for reducing ammonia emissions and in a Guidance

Box 2.1. The Gothenburg protocol

The Gothenburg Protocol sets emission ceilings for 2010 for four pollutants: sulphur, NO_x, VOCs and ammonia. These ceilings were negotiated on the basis of scientific assessments of pollution effects and abatement options. Parties whose emissions have a more severe environmental or health impact and whose emissions are relatively cheap to reduce will have to make the biggest cuts. Once the Protocol is fully implemented, Europe's sulphur emissions should be cut by at least 63%, its NO_x emissions by 41%, its VOC emissions by 40% and its ammonia emissions by 17% compared to 1990.

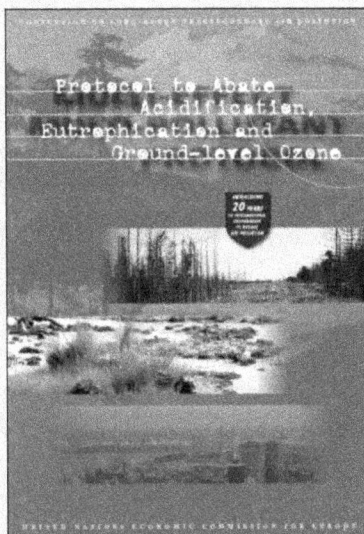

The Protocol also sets tight limit values for specific emission sources (e.g. combustion plant, electricity production, dry cleaning, cars and lorries) and requires best available techniques to be used to keep emissions down. VOC emissions from such products as paints or aerosols will also have to be cut. Finally, farmers will have to take specific measures to control ammonia emissions. Guidance documents adopted together with the Protocol provide a wide range of abatement techniques and economic instruments for the reduction of emissions in the relevant sectors, including transport.

It has been estimated that once the Protocol is implemented, the area in Europe with excessive levels of acidification will shrink from 93 million hectares in 1990 to 15 million hectares. That with excessive levels of eutrophication will fall from 165 million hectares in 1990 to 108 million hectares. The number of days with excessive ozone levels will be halved. Consequently, it is estimated that life-years lost as a result of the chronic effects of ozone exposure will be about 2,300,000 lower in 2010 than in 1990, and there will be approximately 47,500 fewer premature deaths resulting from ozone and particulate matter in the air. The exposure of vegetation to excessive ozone levels will be 44% down on 1990.

Source and more information is available via
http://www.unece.org/env/lrtap/multi_h1.htm

Document to the 1999 Gothenburg Protocol describing abatement techniques. Dutch scientists participate actively in this expert group.

For the European Union Member States EU Directive 2001/81/EC sets upper limits for each Member State for the four components responsible for acidification, eutrophication and ground-level ozone. This is the so-called National Emission Ceilings Directive (NEC) and more specific information is given in the Box 2.2. The 2010 ammonia emission ceiling in the NEC Directive is set to 128 million kg ammonia for all Dutch sectors, similar to the UNECE.

The purpose of the EU Directive 96/61/EC is to achieve integrated prevention and pollution control (IPPC) in different economic sectors, leading to a high

---

Box 2.2. Directive 2001/81/EC of the European Parliament and of the Council of 23 October 2001 on national emission ceilings for certain atmospheric pollutants.

Article 4 of the National Emission Ceilings Directive (NEC) is the most important implementation article of all, because this article determines that by the year 2010 and in later years, Member States shall not exceed their national emissions ceilings as laid down in Annex I of the directive. All the other implementation provisions in fact serve this main objective for which the Member States are responsible by implementing appropriate measures to comply with the national emission ceilings. It was judged that evaluation of the progress made towards compliance with the emission ceilings, would be necessary. Member States have two obligations of this kind.

The purpose of the emission ceilings is broadly to meet the following interim environmental objectives:
• the areas with critical loads of acid depositions will be reduced by at least 50% compared to 1990;
• ground-level ozone loads above the critical level for human health will be reduced by two-thirds compared with the 1990 situation. An absolute limit is also set. The guide value set by the World Health Organisation may not be exceeded on more than 20 days a year; and ground-level ozone loads above the critical level for crops and semi-natural vegetation will be reduced by one-third compared with 1990. An absolute limit is also set.

Source and more information available via http://ec.europa.eu/environment/air/ implem_nec_directive.htm

---

level of protection of the environment as a whole. Intensive livestock farms with more than 40,000 places for poultry, or 2,000 places for production pigs over 30 kg, or 750 places for sows, will need an operating permit. The IPPC technical working group for intensive livestock with substantial input from Dutch agricultural scientists produced an extensive reference report on best available techniques for emission reduction (Anonymous, 2003). Additional information on IPPC is available via internet http://ec.europa. eu/environment/ippc/.

References

Anonymous, 1993. Beoordelingsrichtlijn in het kader van Groen Label stallen. LNV / VROM Report LNV / VROM, The Hague, The Netherlands.

Anonymous, 2003. Reference document on best available techniques for intensive rearing of poultry and pigs. BREF-ILF Report IPPC. European Commission.

Anonymous, 2005. EMEP/CORINAIR Emission inventory guidebook - 2005. EEA Report 30. European Environment Agency, Copenhagen, Denmark.

Bussink, D.W., 1996. Ammonia volatilization from intensively managed dairy pastures. Thesis Wageningen University, Wageningen, The Netherlands, 177 pp.

De Bode, M.J.C., 1991. Odour and ammonia emissions from manure storages. In: Odour and ammonia emission from livestock farming. V.C. Nielsen, J.H. Voorburg and P. L' Hermite, Eds.Elsevier Applied Science, London and New York. pp. 59-66.

De Winkel, K., 1988. Ammoniak-emissiefactoren voor de veehouderij. Publikatiereeks Lucht Report 76. VROM, Leidschendam, The Netherlands.

Henkens, C.H., 1975. The borderline between application and dumping of organic manure (in Dutch). Bedrijfsontwikkeling 6: 247-250.

Huisman, L.H., 1973. Nabeschouwing en conclusies van het Megista-congres. Bedrijfsontwikkeling 4: 654, 659, 660.

Kolenbrander, G.J. and L.C.N. De La Lande Cremer, 1967. Stalmest en gier. Waarde en mogelijkheden. H. Veenman & Zonen N.V., Wageningen, The Netherlands.

Sluijsmans, C.M.J. and G.J. Kolenbrander, 1977. The significance of animal manure as a source of nitrogen in soils. International seminar soil environment and fertility management in intensive agriculture, Tokyo, Japan.

Ten Have, P.J.W., 1971. Ervaringen met zuiveringsinstallaties voor mest en gier. H_2O 4: 98-103.

Van Beek, C.G.E.M., D. Van der Kooij and P.C. Noordam, 1984. Nitraat en drinkwatervoorziening. Report Mededeling 84. KIWA, Nieuwegein, The Netherlands.

Van Breemen, N., P.A. Burrough, E.J. Velthorst, H.F. Van Dobben, T. De Wit, T.B. Ridder and H.F.R. Reijnders, 1982. Soil acidification from atmospheric ammonium sulphate in forest canopy throughfall. Nature 299: 548-550.

Van der Eerden, L.J.M., H. Harssema and J.V. Klarenbeek, 1981. Stallucht en planten. IPO Report R-245. Instituut voor plantenziektenkundig onderzoek, Wageningen, The Netherlands.

Van der Hoek, K.W., 1987. Fosfaatproduktienormen voor rundvee, varkens, kippen en kalkoenen. Report CAD voor bodem-, water- en bemestingszaken in de veehouderij, Wageningen, The Netherlands.

Van der Hoek, K.W., 1994. Method for calculation of ammonia emission in The Netherlands for the years 1990, 1991 and 1992 (in Dutch). RIVM Report 773004003. RIVM, Bilthoven, The Netherlands.

Van der Hoek, K.W., 1998. Estimating ammonia emission factors in Europe: Summary of the work of the UNECE ammonia expert panel. Atmospheric environment 32: 315-316.

Van der Hoek, K.W., 2002a. Input variables for manure and ammonia data in the Environmental Balance 1999 and 2000 (in Dutch). RIVM Report 773004012. RIVM, Bilthoven, The Netherlands.

Van der Hoek, K.W., 2002b. Input variables for manure and ammonia data in the Environmental Balance 2001 and 2002 including dataset agricultural emissions 1980-2001 (in Dutch). RIVM Report 773004013. RIVM, Bilthoven, The Netherlands.

Van Geelen, M.A. and K.W. Van der Hoek, 1982. Stankbestrijdingstechnieken voor stallen in de intensieve veehouderij. Also available as translation 535, National institute of agricultural engineering, Silsoe, UK. Report 167. IMAG, Wageningen, The Netherlands.

Van Ouwerkerk, E.N.J., 1993. Meetmethoden NH_3-emissie uit stallen. Onderzoek inzake de mest en ammoniakproblematiek in de veehouderij 16. Report DLO, Wageningen, The Netherlands.

3. Emission sources, their mechanisms and characterisation

Dick A.J. Starmans

Emission of gaseous compounds can be triggered in a number of ways. In this chapter the different types of emission are characterised with respect to their origin, in order to be able to understand the emission mechanisms. Comprehension of the latter is the cornerstone in the process of understanding how these gaseous emissions should be reduced.

After a general inventory of emission types, this chapter will continue with a condensed view on the mechanisms and key parameters of the emission of volatile compounds. Finally, the link between emission sources and emission factors is discussed.

3.1. Inventory of types of emission

3.1.1. Static emissions

Emissions can be called static when the emission rate is not influenced by the emission itself. Static emissions of compounds can occur from liquid manure surfaces, where the abundance of the compound in the liquid state is hardly influenced by the disappearance of the compound to the gaseous phase. The emissions from a pit filled with slurry can be considered to be static, moreover because of the continuous addition of fresh animal manure from voiding animals.

3.1.2. Reactive emissions

In some cases, chemical reactions can be the source of emissions. Examples of these reactions are decomposition reactions, combustion reactions, enzymatic reactions or electrochemical (half) reactions. In the field of agriculture one should particularly notice the enzymatic breakdown of urea to ammonia and carbon dioxide by urease:

$$H_2N\text{-}CO\text{-}NH_2 + H_2O \longrightarrow 2\,NH_3 + CO_2$$

This reaction occurs rapidly when urine (containing urea) is brought into contact with faeces (containing urease).

3.1.3. Transient emissions

The third category of emissions is the type of emissions that take place only during a relatively short period. These emissions are often triggered by human interaction with a system and only occur during this interaction. After completion of the human interaction, the system normally returns to its former state, hence the term transient.

Often transient emissions are triggered by normal management activities. Typical examples are the periods in which the animals are fed. The change in animal behaviour (i.e. increased voiding during this activity period) can have a great impact on the level and the pattern of emissions (Groenestein *et al.*, 2003). Flushing of a pit, pumping manure, or processing liquid manure all have a distinct influence on the observed emissions.

Also a lack of manure management can cause a transient emission. The emissions of a stagnant pit can be considered transient because of the depletion of the top manure layer (De Bode, 1991). Such circumstances occur for example in outside manure storages where manure is stored for a prolonged period of time without mixing or further manure additions. It also occurs in manure pits in cattle houses, when the cattle is allowed to graze.

3.2. Mechanisms and key parameters of emission of volatile compounds

3.2.1. Static emissions

In general, liquid and gaseous phases are in equilibrium. The concentration of a compound A in the gaseous phase is proportional to the concentration in the liquid phase as predicted by Henry's law (Equation 1).

$$[Compound\ A]_{liquid} = H \times [Compound\ A]_{gas} \tag{1}$$

With: H = Henry coefficient (-)

In extreme cases (i.e. when the gaseous phase is replaced continuously at a great pace), the transport of compound A from the bulk fluid phase to an imaginary film layer on the fluid-gaseous interface can be diffusion limited (Figure 3.1). If this is the case, the concentration of compound A in the film is lower than that of the bulk liquid phase, which results in a decreased overall transport of compound A from the liquid state to the gaseous phase. Henry's law is still applicable, being it with the lower concentration of compound A in the liquid phase that is in equilibrium with a new gaseous bulk concentration.

When considering a liquid phase containing ammonia, the concentration of free ammonia in solution is governed by equilibrium reaction 2.

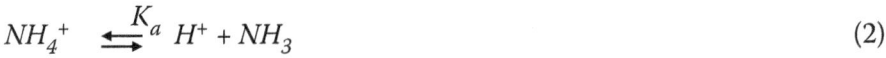

$$NH_4^+ \xrightleftharpoons{K_a} H^+ + NH_3 \tag{2}$$

With: $K_a = \dfrac{[H^+].[NH_3]}{[NH_4^+]} = Acid\ constant\ (mol/l)$

The fraction of unprotonated ammonia in solution is defined in Equation 3, which is rewritten using the acid constant of Equation 2.

$$F \equiv \frac{[NH_3]}{[NH_3] + [NH_4^+]} = \frac{1}{1 + \dfrac{[H^+]}{K_a}} \tag{3}$$

The decrease of the ammonia concentration in time can be described by Equation 4. This equation is valid for ammonia concentrations lower than 45 g/l at T=10 °C and for concentrations lower than 28 mg/l at T=30 °C. In these intervals, Henry's law was found to apply, thus enabling the use of Equation 5.

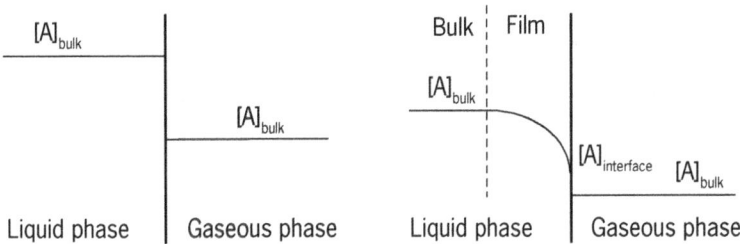

Figure 3.1. Concentration profiles near a liquid-gaseous interface under equilibrium (left) and non-equilibrium circumstances (right).

$$C(t) = C(0) \times e^{-Kt} \tag{4}$$

With: $C(t)$ = the concentration at a certain time t, given in (mg/l)
          K = the desorption rate constant (s^{-1})
          t = the time (s)

The ammonia desorption rate constant K was calculated from literature (Hashimoto and Ludington, 1971).

$$K = A^{-1} \times D \times F \times H^{-1} \tag{5}$$

With: A = the surface area (m^2)
       D = the diffusion coefficient (m^2/s)
       F = the fraction of unprotonated ammonia (-)
       H = the Henry constant (-)

Key parameter in the static emission of volatile compounds is the surface area available for the volatilisation of compound A. Under similar other conditions, an area that has twice the size, can emit twice the amount of a certain volatile compound. This is why many animal houses, especially pig houses, tend to have a reduced pit manure surface area by either designating only part of the animal living quarters for voiding (i.e. application of slatted floor next to a non-slatted lying area (Aarnink *et al.*, 1996)), or by placing inclined walls in the pit. As an example, the latter can – in combination with regular draining of the pit – be used to decrease the ammonia volatilisation by a factor of 4.

Another important parameter is temperature. The temperature dependency of K is found in the definition of the diffusion coefficient D and the acid equilibrium constant K_a from Equation 2. An increase in temperature causes a rise in D and a shift of equilibrium 2 to the right (increase in F). Model calculations showed that for houses for growing-finishing pigs, in the range of 15 to 25 °C, for every degree Celsius increase in floor temperature ammonia emission increased with 1.2%. For every degree Celsius increase in temperature of the manure, ammonia emission increased with 6.8% (Aarnink and Elzing, 1998).

3.2.2. Reactive emissions

Certain emissions occur as a result of chemical reactions. In general, these reactions require certain conditions to be met. Important parameters include temperature and the concentration of reactants, although in a number of instances – i.e. biological reactions – additional conditions such as pH, inhibitor concentration, removal of reaction products from active sites, and the addition of catalysts may influence the overall reaction speed.

With respect to ammonia emissions from cattle and pig houses, the urease activity plays an important role in the ammonia emission rate. Generally, urease activity in the manure pit is not the limiting factor for ammonia generation. The urease activity from the floor however can be limiting in case of metal slatted floors (Aarnink and Elzing, 1998; Aarnink *et al.*, 1997).

3.2.3. Transient emissions

In some static situations, the source of emission is depleted over time. This is the case in non-mixed storage systems, where the amount of volatiles emitted is mainly a function of the surface area (De Bode, 1991). The strong mass transport resistance inside the manure phase is effectively hindering transport to the manure-air surface area.

However, most transient emissions are non-static and often a result of external interaction with a system. The severity of this interaction determines the net effect in terms of temporarily enhanced emissions. In animal care, the day to day routines in management may gradually result in smaller interaction levels caused by adaptation of the animals.

3.3. Determining emission factors

Emission factors are the simplified link between a given practical situation and related emission values. It is possible to have emission factors for specific animal categories that are kept in specific animal houses. Another type of emission factor is that for specific manure application procedures. Both types of emission factors represent essential knowledge for the estimation of gaseous emissions from agriculture to the environment.

From measurements at farms, during certain periods of time, an average emission value can be obtained. The average number of animals present during this time can be used to calculate an emission factor for this animal category that is housed in this specific housing system under the specific management during the measurements. These emission factors are given in grams emitted gas per animal per day. Finally, the obtained values are interpreted by a technical advisory committee (TAC-Rav), to fix an emission factor.

Animal categories are not limited to a certain species. Within a species, one can discern emission-wise between age and sex of the animals. Relevant animal categories have been listed and their respective emissions have been monitored in a range of animal keeping systems, which lead to the compilation of the national list of animal-based emission factors for different types of animal houses in The Netherlands (VROM, 2006).

References

Aarnink, A.J.A. and A. Elzing, 1998. Dynamic model for ammonia volatilization in housing with partially slatted floors, for fattening pigs. Livestock production science 53: 153-169.

Aarnink, A.J.A., D. Swierstra, A.J. Van den Berg and L. Speelman, 1997. Effect of type of slatted floor and degree of fouling of solid floor on ammonia emission rates from fattening piggeries. Journal of agricultural engineering research 66: 93-102.

Aarnink, A.J.A., A.J. Van den Berg, A. Keen, P. Hoeksma and M.W.A. Verstegen, 1996. Effect of slatted floor area on ammonia emission and on the excretory and lying behaviour of growing pigs. Journal of agricultural engineering research 64: 299-310.

De Bode, M.J.C., 1991. Odour and ammonia emissions from manure storages. In: Odour and ammonia emission from livestock farming. V.C. Nielsen, J.H. Voorburg and P. L' Hermite, Eds. Elsevier Applied Science, London and New York. pp. 59-66.

Groenestein, C.M., M.M.W.B. Hendriks and L.A. Den Hartog, 2003. Effect of feeding schedule on ammonia emission from individual and group-housing systems for sows. Biosystems engineering 85: 79-85.

Hashimoto, A.G. and D.C. Ludington, 1971. Ammonia desorption from concentrated chicken manure slurries. Livestock waste management and pollution abatement, St. Joseph, Michigan, USA, ASAE.

VROM, 2006. Regeling ammoniak en veehouderij. Staatscourant 207 (October).

4. Emission inventories

Harry H. Luesink and Gideon Kruseman

Performing a nation wide ammonia emission inventory involves basic processes such as data gathering, interpretation and modeling. In this chapter, the current methodology of performing the Dutch ammonia emission inventory is highlighted with regard to these aspects.

Following the general process and organisation of gathering the necessary data, this chapter will continue with a brief overview of the development of the calculation models used. The scientific assumptions behind the most current model precede the model description and a description of the required data. Finally, calibration and calculations are described, including the most recent results of the Dutch ammonia emission inventory.

4.1. Introduction

In the second national environmental outlook (RIVM, 1991) the national ammonia emissions were estimated to be 251 million kg in 1980, 285 million kg in 1986 and 234 million kg in 1989. In that same report a forecast was made for the ammonia emissions in The Netherlands. National ammonia emissions of 152, 114 and 104 million kg were predicted for the years 1994, 2000 and 2010 respectively. The high emission levels and the required reduction to meet emission goals in the future triggered the yearly calculation of the ammonia emissions in The Netherlands (Brouwer *et al.*, 2002; MNP, 2006b; RIVM, 1995).

Calculation and publishing of the ammonia emission was the task of two institutes: the Agriculture economics research institute (LEI) and the National institute for public health and the environment (RIVM). The Netherlands environmental assessment agency (MNP) emerged in 2005 from the environmental part of RIVM. LEI is concerned with the gathering of activity data related to agricultural sources and the calculation of the ammonia emissions from agriculture. RIVM (nowadays MNP) has a coordination role, calculates emission factors as well and publishes the results and the background information (Hoogervorst, 1991; Van der Hoek, 1994, 2002a,b).

4.2. Process and organisation

Calculation of the ammonia emissions of a certain year is performed in the two following years. At first, a preliminary amount is calculated, and a year later when all activity data are available, the final amount is calculated. The final ammonia emissions are debated in Dutch parliament in the second half of May to evaluate the results of environmental policy and to debate the necessary policy for the coming year. The time schedule of the process and organisation (Table 4.1) is secondary to these discussions.

The Working group Emission Monitoring (WEM) coordinates, controls planning and advises on proposed monitoring programs for a great number of emissions from all agricultural and non agricultural sources. For ammonia emissions the WEM delegates their responsibility to the working group agriculture and land use. This working group holds experts from MNP, Wageningen UR institutes LEI and Alterra, the Institute for inland water management and waste water treatment (RIZA) and Statistics Netherlands (CBS).

The ministries of agriculture and environment not only fund the calculations, but also formalise both calculation methods and results issued by the WEM. MNP decides in September with the advice of the working group agriculture and land use on the proposals for improvement of the calculations.

Not only improving the way that national ammonia emissions are calculated is a time consuming process. Improving the collection of the necessary or new data is a process of three to four years. In this process, researchers

Table 4.1. Present process and organisation of the yearly calculation of the Dutch ammonia emission.

Month	Organisation	Kind of work
June/August	MNP/LEI/others	proposals for improvement
September	WEM/MNP	advice and decisions
October/May	MNP	coordination
October/December	LEI/MNP	collecting data
January	LEI	calculation with MAM/MAMBO
February/May	MNP/LEI	report of results

from LEI and MNP suggest proposals to the different groups who decide if improvements or new data should be adopted.

Besides the organisational efforts for the ammonia emission calculation, the actual data gathering process involves many more parties to ensure timely availability and quality of the data at hand:
1. The annual agriculture census. Proposals for additional (other than yearly gathered) data gathering are made each year in April by the research institutes, CBS and the ministry of LNV. The working group 'Inventory added to annual agricultural census' decides which data is to be collected in the following year(s). It includes data about animal housing, manure storage facilities and manure application systems.
2. The excretion per animal type is updated each year by the working group uniform mineral- and manure excretions. Each September this group evaluates new proposals for these excretion values and improvements of the calculations (Van Bruggen, 2006).
3. Data from the Dutch Farm Accountancy Data Network (BIN). At the end of each year the management of LEI evaluates proposals considering the data that would be collected in the following year. Regarding ammonia emissions, data on grazing time and grazing systems for cattle and application amounts of manure per crop are of importance.
4. Data that are used to control execution of the manure laws. The ministry of agriculture collects these data and decides which part could be given to research institutes. These data concern transport, export and processing of manure, in combination with application norms and soil types.
5. Ammonia emission factors. MNP did this work until 2004. To continue the updating of this data the ministry of agriculture installed a working group in 2006.

4.3. Development of calculation models

Already in 1982 LEI started with the constructing of a model for the calculation of technical and economical aspects of manure distribution and processing (Manure model). Financed by the precursor of FOMA (see Chapter 1.3) 'Commission on prevention of nuisance from livestock farms' this research yielded the first model in 1984 (Wijnands and Luesink, 1984).

With research guided by FOMA at the end of the 1980's an ammonia emission model was constructed by the LEI (Oudendag and Wijnands,

1989). In that year also the second manure model was finished (Luesink and Van der Veen, 1989). In the beginning of the 1990 both models were combined to the first LEI manure and ammonia model called *MestAmm* (Brouwer *et al.*, 2001; Oudendag and Luesink, 1998).

In 1996 LEI started with the construction of the second generation of the Manure and ammonia model (MAM). It included variant and version management and was finished in 1998 (Groenwold *et al.*, 2002; Helming *et al.*, 2005).

Late 2004, LEI started with the third generation of the manure and ammonia model for policy support (MAMBO). This generation was finished in 2006. It allowed the scientific decoupling of emission data and their origin, thus allowing the calculation of the national emission values using local originated data.

It is eminent that in the near future, the gathering of data sources and application of calculation protocols regarding ammonia emissions will receive international attention. Harmonisation of the complete process should become an integral part of the EU agenda.

4.4. The scientific base of MAM/MAMBO

By the development of MAMBO, a generic formulation was chosen to facilitate the use of data with a deviating structure (i.e. animal categories, crops, manure categories, housing types). Furthermore, adjustments to incorporate the policy concerning manure and emissions in MAMBO were made. The generic formulation ensures backward compatibility with the MAM model.

MAMBO can be used to calculate both nutrient flows and ammonia emissions (Figure 4.1). To establish this, data on five key processes regarding animal manure are gathered and processed in this model:
1. manure production on farm;
2. on farm maximum allowed application of manure within statutory and farm level constraints;
3. manure excess at farm level (production minus maximum application amount);
4. manure distribution between farms (transport);
5. soil loads with minerals.

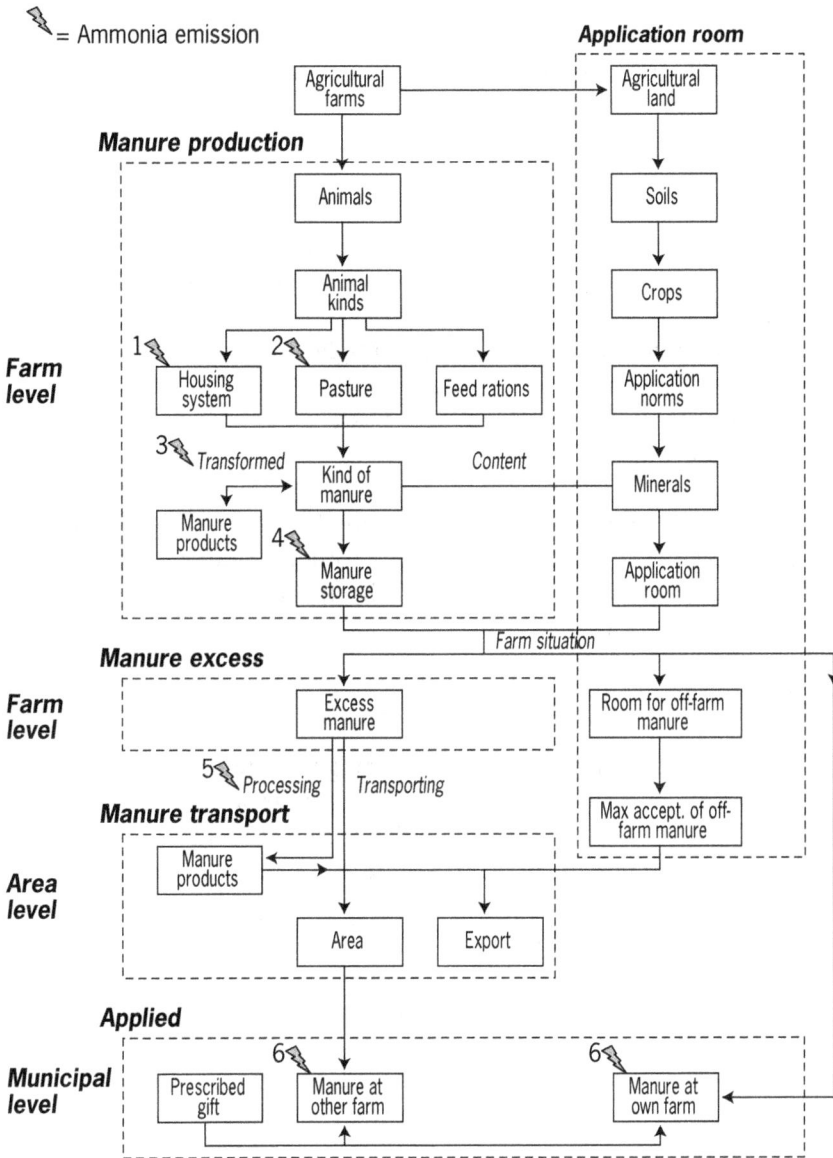

Figure 4.1. The Manure and Ammonia emission Model (MAM/MAMBO).

The calculations take place at three spatial levels. The first three processes are calculated at farm level, whereas manure distribution is calculated at the level of 31 predefined manure regions, and soil loads are calculated at municipality level. These five key processes are described in further detail, prior to dealing with ammonia emissions on the basis of the three spatial levels in the next part of this chapter.

4.4.1. Manure production

Manure produced on animal farms can be classified and processed separately in the MAMBO model. Sources of manure are distinguished based on the following parameters:
1. type and number of animals kept on the farm;
2. type of feed given to the animals;
3. housing facility (yes – housed, no – pasture);
4. type of housing facility used.

The manure can be excreted directly on the field, it can be stored or it can be processed at farm level into other products, such as dried manure or separation products, each with its specific ammonia emission characteristics.

4.4.2. Maximum application amount

MAMBO includes three factors determining the amount of manure for the application of on-farm manure: the total crop area (including grassland) of the farm, the type of crops (including grass) grown on the farm, and the statutory application standards. The statutory application standards prescribe the maximum amount of nitrogen and phosphate allowed to be applied for each crop and soil type.

A farm with more manure production than its maximum application amount can still accept off-farm manure in cases where the on-farm manure is not suitable or economical for the type of crops grown on the farm. A larger part of the on-farm produced manure then has to be transferred to other farms to avoid surpluses.

4.4.3. Manure excess at farm level

There are several ways in which manure, either processed or unprocessed, can be used. It can be applied on the land of the farm where it is produced, stored or transported to other farms. Furthermore, there are a number of conditions for the manure production by animals kept on pasture. Firstly, pasture (grassland) needs to be part of the cropping plan of the farm. Secondly, manure from pasture can neither be transported nor processed. Thirdly, the manure production from pasture may not exceed the statutory application norms for grassland of the particular farm.

In order to determine whether a farm has a manure surplus or room for off-farm manure, the manure produced on the farm is balanced against the maximum application amount of manure on the farm. In a case of manure surplus, the economic consequences of the surplus are minimised by finding the most appropriate type of manure for each particular farm.

The maximum amount of off-farm manure applicable on a farm depends on the farmer's willingness to accept off-farm manure and on the actual maximum application amount. In normal life, this is determined by the nutrient requirements of the crops grown on the farm, the region and the price of manure. In MAMBO, the willingness to accept off-farm manure is translated to maximise crop returns.

4.4.4. Manure transport

MAMBO includes three options for manure that cannot be applied at farm level: it can be transported to other farms within the same region, other regions or exported to other countries, either processed or unprocessed. Given the necessity for a farm to transport manure, the main driver for transport of any type of manure is minimising manure transfer costs. Receiving farms aim to maximise crop returns.

The combined data on farm total manure surplus, total application amount for off-farm manure, and the available options for manure processing and export, is used in the MAMBO model to calculate manure transfers within and between 31 predefined regions. The transfers are calculated in such a way that costs are minimised at national level. The costs of transport, storage, application, processing and export are deducted from the revenues of the manure sales.

Whether manure is transferred within the same region, to other regions, or exported depends on the transportation costs, the expected revenues of the manure and the maximum application amount for off-farm manure. Transportation costs within a region are fixed and depend on the type of manure and the type of application. Transport between regions is also dependent on the distance between the regions. Manure revenues are measured on the basis of its fertilisation value. The fertilisation value is determined for each crop on the basis of the following characteristics: (1) the contents of nitrogen, phosphate and potassium, (2) the fertilisation advice for the particular crop, (3) the content of unwanted substances, such as copper or weed seeds, and (4) the manageability of the manure type, with processed manure being more manageable and valuable than unprocessed manure.

Transport costs are minimised within the scope of these basic assumptions:
1. Processing and export of manure may not exceed maximum capacities.
2. Regional manure mass balance: The sum of the total manure production of a region and the supply of manure from other regions must be equal to the sum of regional application of manure, off-farm manure and processing minus export and transport to other regions.
3. The manure transport into any region is equal or less than the available room for off-farm manure for that region.
4. Manure is transferred from other regions only if the regional surpluses are insufficient to fill up the room for off-farm manure.
5. Manure is transported into other regions only if it is in surplus, exceeding the maximum application amount for off-farm manure in the region of origin.

4.4.5. Soil loads with minerals

In MAMBO, the total mineral load of the soil depends on three factors: the application of on-farm manure, the application of off-farm manure and the application of mineral fertiliser. The Dutch Farm Accountancy Data network provides data and statistics available about the use of mineral fertilisers at a regional level. These are divided at municipal level with a distributive code. The distributive code holds data on the time of manure application, the effectiveness of the nutrients and the amount of nutrients in the applied manure. For this purpose, the manure transfers on municipality level are calculated from the results of manure transfers on regional level by disaggregating these to municipality level.

4.5. Mathematical representation of ammonia emissions in MAMBO

MAMBO is a suite of modules written in GAMS (General algebraic modelling system (McCarl, 2006)). MAMBO follows a modular approach and allows for calculations at varying levels of aggregation depending on the availability of data. The lowest levels that MAMBO takes into consideration are individual animals, plots, stables, complete industrial processes (i.e. for manure processing). It also accounts for interactions between parties handling manure through a spatial equilibrium model where suppliers of manure, i.e. livestock farmers with a surplus amount meet arable farmers with ample space for manure placement. The model calculates the transport of manure at municipal level and finally the placement of manure and additional artificial fertiliser are calculated at plot level on the farm.

The structure of MAMBO allows for calculations to take place at a higher level of aggregation if the available data and / or the policy or research question at hand deem it more appropriate. In this section the mathematical structure of relevant parts of the model that attribute to the calculation of ammonia emissions is presented.

The mathematical representation of the model equations follows the standards presented in Table 4.2 (indices). The indices represent a subdivision of the variable at stake. These subdivisions are related to the identification of the variable (e.g. region, municipality, farm, establishment) or a further specification of the variable it self (e.g. manure categories, crops, derogation). They are presented in alphabetical order.

In the first calculation module, animal numbers are converted into manure quantities by taking into account the housing situation of the animals and whether or not they are grazing in summer. The common housing and grazing circumstances (mathematically expressed as departments with each a certain emission characteristic) are obtained from the annual agriculture census and the Dutch Farm Accountancy Network described earlier.

Manure production per animal category (Q^{animal}). The manure production depends on the number of animals, the ration (ρ) the animals are fed, the excretion volume (v) of the animal and the department in which the animal is located. Rations are considered equal when an animal is housed indoors or outdoors. The department is in general an animal house. Time fraction (τ) is used to assign more than one department (pasture in summer and stable in winter) to animals during a year, where relevant. The dimension is kg manure per animal category per department per farm establishment.

$$Q^{animal}_{FERrm,\,\rho\delta Da} = N^{animals}_{FERrm,\,\delta Da} \times \rho_{FERrm,\,\rho a} \times v_{\rho a} \times \tau_{FERrm,\,\rho\delta da} \tag{1}$$

Table 4.2. Indices in MAMBO.

Symbol	Explanation	Symbol	Explanation
A	manure aspect type	m, n, o	municipality
a	animal categories	R	large regions
C	broad groups of crops	r, q, p	smaller regions
c	crop type	S	storage category
D	category of places where animals stay (housing/pastures)	s	sector level destination of surplus fertiliser
d	derogation (yes, no)	w	application method
E	farm establishment		
F	farm identification		
f	specific field or plot	ε	emission factors
G	geographic destination of surplus fertiliser	δ	specific place where animals stay (housing/pastures)
H	manure source (own, foreign)	ϕ	broad groups of fertiliser with a similar origin
J	manure factories	φ	manure and fertiliser categories
K	mineral fractions in terms of effect	γ	country
L	process to convert manure	ρ	feed rations or menus
M	mineral type	σ	soil type

Within MAMBO, manure categories are defined in terms of the animals that produce the manure and the departments where the manure is produced.

$$Q^{animal}_{FERrm,\, \rho\delta Da\varphi} \Leftrightarrow Q^{animal}_{FERrm,\, \rho\delta Da} \tag{2}$$

Mineral production (M^{animal}) of an animal in a department for a manure category depends on the mineral content of the manure excreted (μ). The dimension is kg mineral in manure per animal category per department (hence per mineral category) per farm.

$$M^{animal}_{FERrm,\, \delta Da\varphi,\, M} = \sum_{\rho}(Q^{animal}_{FERrm,\, \rho\delta Da\varphi}) \times \mu_{M\varphi} \tag{3}$$

The emission factor for grazing ($\varepsilon^{pasture}$) is different from that of the animal house (ε^{stable}). Hence, the mineral emissions (E) from the animal manure inside the animal house and from grazing are expressed separately in

Equations 4 and 5. The emission is expressed as kg mineral emitted per animal category per department (hence per mineral category) per farm and emission kind (one of them is ammonia).

$$E^{stable}_{FERrm,\ \delta Da\varphi,\ M\varepsilon} = M^{animal}_{FERrm,\ \delta Da\varphi,\ M} \times \varepsilon^{stable}_{\varepsilon pD} \qquad \text{(4, flag 1 in Figure 4.1)}$$

$$E^{pasture}_{FERrm,\ \delta Da\varphi,\ M\varepsilon} = M^{animal}_{FERrm,\ \delta Da\varphi,\ M} \times \varepsilon^{pasture}_{\varepsilon pD} \qquad \text{(5, flag 2 in Figure 4.1)}$$

The mineral production per animal after stable and pasture emission is calculated by adding up the two emission variables. The mineral production (M) after emissions of minerals at animal level is given in Equation 6.

$$M^{animal\ after\ emissions}_{FERrm,\ \delta Da\varphi,\ M} = M^{animal}_{FERrm,\ \delta Da\varphi,\ M} - E^{pasture}_{FERrm,\ \delta Da\varphi,\ M\varepsilon} - E^{stable}_{FERrm,\ \delta Da\varphi,\ M\varepsilon} \qquad (6)$$

The emission residual (R) at animal level is defined in Equation (7). The emission residual is an endogenous correction factor on the mineral manure content.

$$R^{animal\ after\ emissions}_{FERrm,\ \delta Da\varphi,\ M} = \frac{M^{animal\ after\ emissions}_{FERrm,\ \delta Da\varphi,\ M}}{M^{animal}_{FERrm,\ \delta Da\varphi,\ M}} \qquad (7)$$

Manure processing on a farm level is not yet implemented in MAMBO. The calculation rules in MAMBO for ammonia emissions from manure processing systems at farm level (flag 3 in Figure 4.1) are expected to be incorporated in 2007.

Emissions from manure storage at farm level are calculated at department level in the Aggregate Manure Production Calculations module.

$$E^{storage}_{FERrm,\ \delta D\varphi,\ M\varepsilon} = S_{FERrm,\ Ds} \times \varepsilon^{storage}_{\varepsilon Ds} \times \sum_{a} M^{animal\ after\ emissions}_{FERrm,\ \delta Da\varphi,\ M} \qquad \text{(8, flag 4 in Figure 4.1)}$$

Department mineral production and emission residual are calculated in a comparable way to those at animal level. This module also calculates aggregate results at various levels within the farm. Further aggregation to municipal and regional levels is done by summation.

For the national ammonia emission the calculation rules in MAMBO for calculation of the manure excess and manure distribution are of minor importance. In the context of this book they are not important, that's why we don't give a detailed description of it. For a general description see paragraphs 4.4.1 through 4.4.5. In this part of the model the ammonia emission of processing at regional level takes place. As given in Equation 9, the emissions from processing depend on the amount of manure processed, the mineral content of that manure and the way of processing.

$$E^{process}_{\varphi\to\phi,\,M\varepsilon} = \varepsilon^{process}_{\varphi\to\phi,\,M\varepsilon} \times \sum_{Rrm} \left(\mu_{M\varphi} \times \sum_{FE} \left(Q^{process}_{FERrm,\,\varphi} \right) \times R^{average}_{Rrm,\,\varphi,\,M} \right) \qquad \text{(9, flag 5 in Figure 4.1)}$$

Emission from the application of manure is calculated in the following fashion. Emissions depend on the quantity of minerals in manure applied. This is in turn and calculated by multiplying the applied manure volume with the manure mineral content corrected for the emissions at animal, department and storage level.

$$E^{application}_{rm,\,co\varphi M\varepsilon} = \sum_{qn,\,w} Q^{application}_{qn\to rm,\,\varphi c\sigma} \times R^{average}_{qn,\,\varphi M} \times \mu_{M\varphi} \times \varepsilon^{application}_{rwf} \times w_{rm,\,wc\sigma} \times f_{f\varphi}$$

$$\text{(10, flag 6 in Figure 4.1)}$$

Ammonia emissions depend on the application method (w) and on the fraction of mineral nitrogen in the manure ($f_{f\varphi}$). Nitrogen in manure exists of two fractions: mineral nitrogen and organic nitrogen. Finally we calculate the emissions from inorganic fertiliser application. Application of inorganic fertiliser is the result of farm level decision making processes and is correlated with the difference between total organic mineral application to the crops and the mineral application advice for the crop soil type combination. Agricultural household theory (Kruseman, 2001; Singh *et al.*, 1986) dictates that farmers decisions depend on household (ξ), farm (ζ) and institutional characteristics (v).

$$M^{inorganic}_{FERrm,\,M\varphi,\,c\sigma} = g\left(e\left(\overline{M}^{optimal}_{M,\,c\sigma} - \sum_{\phi\neq\varphi}(M^{manure}_{FERrm,\,M\varphi,\,c\sigma}) \right), \xi, \zeta, v \right) \qquad (11)$$

With

$$M^{manure}_{Rrm,\,M\varphi,\,c\sigma} = \mu_{M\varphi} \times \sum_{qn,\,w} \left(Q^{application}_{qn\to rm,\,\varphi c\sigma} \times R^{average}_{qn,\,\varphi M} \right) \varepsilon^{application}_{rwf} - \sum_{\varepsilon} E^{application}_{Rrm,\,co\varphi M\varepsilon} \qquad (12)$$

Theoretically this implies that the decision to apply inorganic fertilisers is taken jointly with the decision to apply organic fertilisers from animal origin. Following the seminal work of Heckman on estimation of farm level model with endogenous variables (Heckman and McCurdy, 1982), we have developed a methodology for disentangling these effects using an econometric model that corrects for partial endogenous parameters. Full explanation of the methodology surpasses the context of this chapter. Interested readers are invited to literature in other contexts (Hagos *et al.*, 2006; Kabubo-Mariara *et al.*, 2006). A description of the model in the context of farm household decisions and emission of pollutants is forthcoming.

In the present context we aggregate from the household level to the municipal level. This has a number of advantages. First of all the errors at the household level tend to be smoothened at the aggregate level leading to aggregation gains, a number of idiosyncratic characteristics of individual farms are also lost. As a result we use an aggregate function calibrated on survey data from BIN, the Dutch farm accountancy data network.

$$M_{Rrm,\,M\varphi,\,co}^{inorganic} = \tilde{g}\left(\left(\overline{M}_{M,\,co}^{optimal} - \sum_{FE\phi \neq \varphi} (M_{FERrm,M\varphi,co}^{manure}) \right), \hat{M}_{Rr,M\varphi,co}^{inorganic} \right) \tag{13}$$

The emissions from the use of inorganic fertiliser are calculated using Equation 14.

$$E_{Rrm,\,Me}^{inorganic} = \varepsilon_{Me}^{inorganic} \times \sum_{\varphi,co} M_{Rrm,M\varphi,co}^{inorganic} \tag{14, flag 6 in Figure 4.1}$$

4.6. Activity data

4.6.1. General

For the calculation of the ammonia emission a lot of data are used. These data differ in detail, frequency of collecting and source (Table 4.3).

The data sources presented in Table 4.3 depend on the level of origin at which the data is collected and the used model (MAM or MAMBO). MAMBO eliminates the influence of origin in such a way that inputs can vary from an individual animal level to national level.

In 2004 the collecting level of housing systems was at farm level, but with MAM these had to be used at a regional level (Table 4.3). With the calculation of the ammonia emission with MAMBO it is possible that all the input data can be used at farm level. Data about grazing systems, housing and grazing period are available at regional level. In both models they can be used at that level.

Table 4.3. Data source, gathering method, frequency of collecting and prior indices to calculate ammonia emissions in The Netherlands.

Emission Type - Parameters	Level[1]	Freq.[2]	Source[3]	Index	Remarks
Manure production					
- Number of animals	F	Y	LBT	animal kind	
- Excretion	N	Y	WUM	ration+mineral	dairy: 4 rations
Emission from housing					
- Housing system	R/F	4Y	LBT	animal kind	dairy: level = f
- Grazing system and housing	N	Y	BIN	animal kind	wum equivalent
period					
- Emission factors	N	Occ.	MNP	housing system	
Emission from grazing					
- Grazing system and period	N	Y	BIN	animal kind	wum equivalent
- Emission factor	N	Occ.	MNP	no index	
Emission from storage					
- Amount of manure	N	Occ.	LBT	manure type	dated
- Storage period	N	Occ.	Research	manure destination	dated
- System	N	Occ.	LBT	manure type	dated
- Emission factors	N	Occ.	MNP	storage system	
Application of manure and fertiliser					
- Crop area	F	Y	LBT	crop kind	
- Legal limits	R	Y	LNV, BIN Research	crop/soil kind, mineral	application gifts
- Distribution between farms	P	Y	LNV	manure kind, mineral	only manure
- Application systems	R	5Y	LBT	manure type	only manure
- Total available N	N	Occ.	Research	manure kind	
- Emission factors	N	Occ.	MNP	manure type, application system	
- Acceptation degree	R	Y	BIN	crop kind	application gifts
- Export	N	Y	LNV	manure type	

[1]Levels are: Farm (F), National (N), Regional (R) and Provinces (P).
[2]Frequencies are: Yearly (Y), 4-Yearly (4Y), 5-Yearly (5Y), Occasionally (Occ.).
[3]Sources are: LBT = annual agriculture census; WUM = working group uniform mineral and manure excretions; BIN = Dutch farm accountancy data network; Research = different kind of research reports; LNV = Ministry of Agriculture, Nature and Food quality; MNP = Netherlands Environment Assessment Agency. In 2006 a working group is installed to calculate the emission factors.

The mineral losses in the manure laws of the mineral accounting system (MINAS) are at national level, whereas the differences in mineral uptake of crops are at regional level. Application limits are the sum of these, which has then a mixed level of origin.

The acceptation degree of manure application, the extent to which limits will be reached, is only for off-farm produced manure. For on-farm produced manure it is assumed that the limits will be filled up. The fertiliser gifts are taken from the Dutch Farm Accountancy Data Network and used as input for the limiting values in the MAM/MAMBO model.

The manure export data are needed to calculate whether ammonia emissions of a certain amount of manure will take place in The Netherlands or abroad.

4.6.2. Housing and storage systems

Nowadays, ammonia emissions from animal housing account for nearly half of the total ammonia emissions from agriculture in The Netherlands. This is recognised in MAMBO and subsequently in the agriculture census that includes the inventory of housing systems used for the most important animal categories every four years. The last inventory of housing systems dates back to 2004 with the upcoming scheduled for 2008. It holds activity data for a number of housing systems:
* *Dairy cows*: ten housing systems (six cubicle housing systems from which two with low ammonia emission; two types of tied housing including one with low ammonia emission; one type of deep litter and one for other housing systems).
* *Dairy calves and heifers*: the same ten housing systems as for dairy cows.
* *Sows*: four housing systems.
* *Fattening pigs*: four housing systems.
* *Piglets*: four housing systems.
* *Laying hens younger than 18 weeks*: seven housing systems (two batteries; two aviary; two ground housing and one other).
* *Laying hens 18 weeks and over*: fourteen housing systems (two batteries with slurry; six batteries dry manure; two aviary, two ground housing and one other).
* *Broilers*: three systems (two traditional and a single low ammonia emission housing).

The results of the inventories of housing systems form the input parameters in the model used to calculate the ammonia emission at regional level. Information on dairy cow housing systems is used at farm level.

As discussed earlier, MAMBO allows the use of all this data in the calculations at farm level. The activity data for pig housing systems with a reduced ammonia emission were taken from the agriculture census of 2001 (Hoogeveen *et al.*, 2005). In that year, 13% of the fattening pigs and 16% of the sows were kept in housing systems with a reduced ammonia system. In 2004, 7% of the dairy cows, 6% of dairy calves and heifers, 9% of laying hens younger than 18 weeks, 21% of laying hens 18 weeks and over and less than 5% of the broilers were kept in housing systems with a reduced ammonia emission.

Updating the activity data on outside storage facilities has received little priority, because the ammonia emission from outside storages has never been higher than 3% of the total ammonia emission. The last inventory of outside storage systems took place in 1997. The results of these inventories at farm level with the agriculture census are still used in the calculation of the Dutch ammonia emission. In 2007 however there will be a new inventory at farm level with the annual agriculture census. The results of this inventory can be used for the calculation of the ammonia emissions at the end of 2007.

4.6.3. Grazing and application

Within the Dutch farm accountancy data network (a sample of about 1500 agriculture and horticulture farms), every year an inventory is made of the grazing systems in use. Starting in 2005, also information about the grazing period is recorded. Each year, this inventory is used to calculate an average situation of the actual grazing systems in The Netherlands. These data are used in further calculations in MAMBO. The distribution of dairy cows per grazing system in 2004 is given below:
• 18% summer feeding in the stable;
• 30% day and night grazing and milking in the parlour;
• 52% limited grazing.

The amount of manure produced in the cattle house during the grazing period depends on the grazing system. Manure production inside the animal house is estimated to be 100%, 15% and 60% respectively for the grazing systems given above.

The last couple of years about 30% of the total ammonia emissions in The Netherlands take place when manure is applied to the land. In this case it is important to know where the manure is applied and with what kind of application technique. The distribution of manure applied on grassland and arable land is updated every year per region using the results of the Dutch farm accountancy data network. In 2004, 61% of the nitrogen in manure was applied on grassland and 39% was applied on arable land.

Approximately once every 5 years, in the national agriculture census, farmers are questioned for information about the kind of manure application techniques they use. In MAMBO, the calculation of the ammonia emission is based on the results of this inventory on a regional scale. The results of the last inventory in 2005 at national average for grassland and arable land are given in Table 4.4 (Hoogeveen *et al.*, 2006).

4.7. Validation and calibration of MAM/MAMBO

To guarantee an accurate result, models need to be validated and calibrated. Over the years, MAM calculated emissions have been validated by measurements in the field (Oudendag, 1999; Smits *et al.*, 2005). It was concluded that emission differences fell within expected margins. However, it was revealed that MAM was sensitive for the level at which housing data was provided. With housing data input at regional level, the ammonia emission was underestimated 15%. This problem was solved by providing

Table 4.4 Distribution of manure application techniques.

Manure application technique	Applied to	Percentage
Closed slot shallow injection and deep injection	grassland	56%
Open slot shallow injection	grassland	14%
Trailing shoe / trailing hose	grassland	23%
Other systems	grassland	7%
Injection	arable land	34%
Trailing shoe / trailing hose	arable land	6%
Surface spread and incorporated in one track	arable land	27%
Surface spread and incorporated in two tracks	arable land	27%
Other systems	arable land	6%

these data on farm level. The difference between calculated and measured emission values proved to be less than 1%. It was also concluded that similar to the housing data, also the data on manure spreading and farm area location should be known at farm level.

In 1999 a group of scientists reviewed the calculation rules and the principles of the calculation of the ammonia emission with MAM (Steenvoorden *et al.*, 1999). They made a couple of recommendations to improve the calculation of the ammonia emission. Most of the recommendations addressed the principles and the available data, not on the calculation rules. In 2004, it was concluded that most of these recommendations were implemented in the calculation methods of the Dutch national ammonia emission inventory (De Mol, 2004).

In 2001 the IT-quality aspects of MAM were audited by a team that reviewed all the models and data files used by MNP (Hordijk, 2004). The audit team concluded that MAM was one of the best models at IT aspects.

Each year, the manure distribution algorithm of MAM is calibrated with statistical data on the transport of manure (Luesink, 2002). As a result of the actual manure legislation, each transport needs a certificate, which is registered to facilitate supervision of the execution of these laws. CBS provides the statistical data on these manure transports to LEI.

4.8. Inventory results

The results of the ammonia emission inventory are published in many documents and publications at different aggregation levels, for instance:
- publications from LEI (Brouwer *et al.*, 2002; Hoogeveen *et al.*, 2006; Luesink, 2004): national and regional results;
- publications from MNP Milieubalans (MNP, 2006b) and Milieucompendium (MNP, 2005): national results; and
- public database of Pollutant Emission Register (ER) (MNP, 2006a): results at a level of 5 * 5 km.

Table 4.5 presents the Dutch ammonia emission from different sources over time. The data given are the official ammonia emissions which The Netherlands reports to the European Union. The emission of housing and storage is added together because manure is mainly stored indoors in The

Table 4.5. Ammonia emission from Dutch agriculture 1980 - 2004 (mln. kg of ammonia). Source: www.emissieregistratie.nl.

	1980	1985	1990	1995	2000	2004
Animal manure	204	227	222	166	128	107
Housing and storage	77	85	87	90	73	59
Grazing	14	16	16	14	10	8
Application	114	126	119	62	45	40
Fertiliser	15	16	13	13	11	13
Total agriculture	220	243	235	179	139	120
Emission per ha agriculture area (kg NH$_3$)	107	118	110	90	71	62
Index (1980 =100)	100	110	107	81	63	55

Emission data are based on Luesink (2004) and Hoogeveen *et al.* (2006).

Netherlands and the emission factors of housing include indoor storage of manure. Only a part of the manure is stored outside the animal houses, in the 80's this part was very small (almost no slurry and about 50% of the solid manure). At the end of the 90's about 50% of cattle manure, 20% of pig manure and almost all solid poultry manure were stored outside the animal house. Due to legislation, all these outside stores had to be covered, and this leads to an emission of 4 million kg of ammonia from outside storage. At that time this value was about 2.5% of the total ammonia emission in The Netherlands.

Nowadays the national ammonia emission is half of the maximum value calculated in 1985. There are a couple of reasons why the ammonia emissions declined:
- Introduction of the milk quota caused a reduction in the number of dairy cattle from 4.2 million heads in 1985 to 2.6 million heads in 2004. The reduction in numbers is caused by an increasing milk yield per cow.
- Laws prescribing manure application techniques with low emission factors are first implemented in 1988 at arable land and in 1991 at grassland. In 1995 they were fully implemented for all areas in The Netherlands.
- Buying of animal production rights by the government in 2001 and 2002 caused a decrease in the amount of pigs and poultry of about 15%.

The last few years the trend of a declining ammonia emission from agriculture has stabilised at around 120 million kg ammonia per year. The ammonia emission from non-agricultural sources in The Netherlands is about 13 million kg. Thus, the total ammonia emission in The Netherlands ranges from about 130 to 135 million kg in the last few years. This is almost the NEC target of 128 million kg in 2010 (MNP, 2006b).

As seen in Table 4.5, the ammonia emission from grazing animals slowly declines over the last few years. Besides the structural decline in the number of grazing animals it also originates from changes in the amount of nitrogen in fed roughage. Due to the Dutch manure laws (MINAS-system) the use of nitrogen fertiliser on grassland declined from more than 250 kg per hectare in 1998 to about 170 kg in 2002 and 2003, which led to a lower nitrogen content in on-farm produced roughage (Luesink and Wisman, 2005). The decline of total ammonia emission would be even more when the grazing systems in the same period did not change from day and night grazing, to more limited grazing and summer feeding. Decreasing the grazing period results in more nitrogen collected in the animal house and the additional ammonia emissions from housing and manure application is more than the ammonia emissions which should have been taken place in the meadow.

Figure 4.2 shows the Dutch ammonia emission from each area of a superimposed 5x5 km grid for the years 1980 and 2002. This figure underlines the sharp decrease in ammonia emissions presented in Table 4.5. It also shows the contours of the three regions with high ammonia emissions, located in the south east, the central east and the central part of The Netherlands.

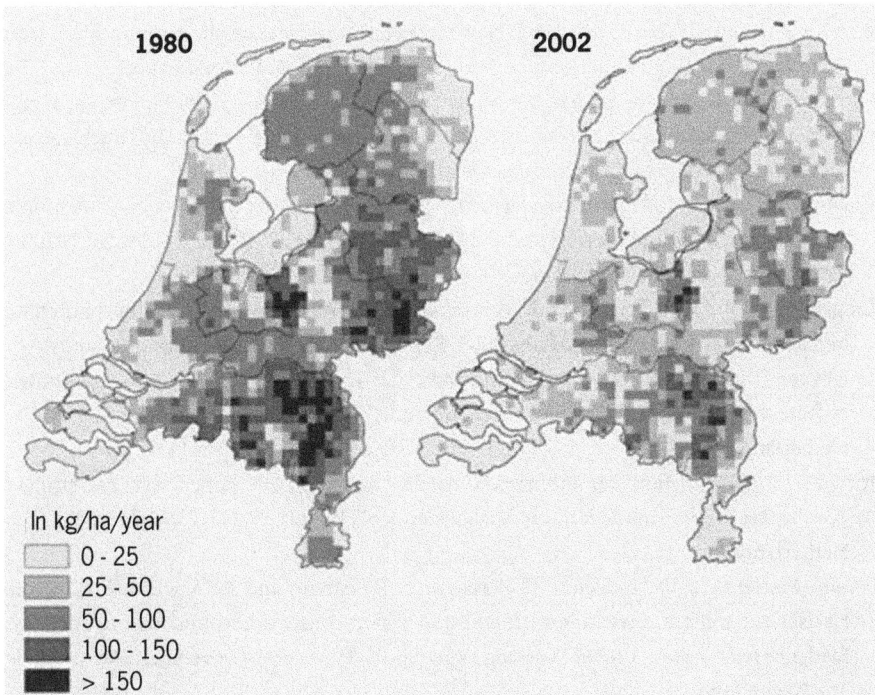

Figure 4.2. Ammonia emission in The Netherlands in kg per ha per year in 1980 and 2002 (RIVM/CBS, 2004).

References

Brouwer, F.M., C.J.A.M. De Bont and C. Van Bruchem, 2002. Landbouw, Milieu, Natuur en Economie. Rapport PR.02.02. LEI, The Hague, The Netherlands.

Brouwer, F.M., P. Hellegers, M.W. Hoogeveen and H.H. Luesink, 2001. Nitrogen pollution control in the European Union: challenging the requirements of the Nitrate Directive with the Agenda 2000 proposals. International journal of agricultural resources, governance and ecology 1: 136-144.

De Mol, R.M., 2004. Evaluatie van de lijst van aanbevelingen in Steenvoorden *et al.* Report Agrotechnology and Food Innovations, Wageningen, The Netherlands.

Groenwold, J.G., D.A. Oudendag, H.H. Luesink, G. Cotteleer and H. Vrolijk, 2002. Het mest- en ammoniakmodel. Report 8.02.03. LEI, The Hague, The Netherlands.

Hagos, F., M. Yohannes, V. Linderhof, G. Kruseman, A. Mulugeta, G.G. Samuel and Z. Abreha, Eds., 2006. Micro water harvesting for climate change mitigation: Trade-offs between health and poverty reduction in northern Ethiopia. PREM Working Paper 06-05. IVM, Vrije Universiteit, Amsterdam, The Netherlands.

Heckman, J.J. and T.E. McCurdy, 1982. New methods for estimating labor supply functions: A survey national bureau of economic research, Inc. NBER Working Papers.

Helming, J.F.M., M.W. Hoogeveen, L.J. Mokveld and H.H. Luesink, 2005. Linking farm and market models to analyse the effects of the EU Nitrate directive for the Dutch agriculture sector. EAAE-congres, Copenhagen, Denmark.

Hoogervorst, N.J.P., 1991. The scenario for Dutch agriculture in the second National environmental outlook; Assumptions and results (in Dutch). RIVM Report 251701005. RIVM, Bilthoven, The Netherlands.

Hoogeveen, M.W., H.H. Luesink, L.J. Mokveld and J.H. Wisman, 2005. Uitgangspunten en berekeningen voor de milieubalans 2005. Report LEI, The Hague, The Netherlands.

Hoogeveen, M.W., H.H. Luesink, L.J. Mokveld and J.H. Wisman, 2006. Uitgangspunten en berekeningen voor de Milieubalans 2006. LEI Report In press. LEI, The Hague, The Netherlands.

Hordijk, L., 2004. Mest- en ammoniak model, audit in het kader van het project 'Kwaliteitsborging modellen en databestanden'. Report WUR, Wageningen, The Netherlands.

Kabubo-Mariara, J., V. Linderhof, G. Kruseman, R. Atieno and G. Mwabu, Eds., 2006. Household welfare, investment in soil and water conservation and tenure security: Evidence from Kenya. PREM Working Paper 06-06. IVM, Vrije Universiteit, Amsterdam, The Netherlands.

Kruseman, G., 2001. Household technology choice and sustainable land use. In: Economic policy reforms and sustainable land use in LDCs: Recent advances in quantitative analysis. N.B.M. Heerink, H. van Keulen and M. Kuiper, Eds. Physica-Verlag, Heidelberg, Germany. pp. 135- 150.

Luesink, H.H., 2002. Acceptatie van mest per gewasgroep in 1996, 1997, 1998 en 1999. Report LEI, The Hague, The Netherlands.

Luesink, H.H., 2004. Ammoniak uit de landbouw verder teruggelopen. Agrimonitor 10 (3): 8.

Luesink, H.H. and M.Q. Van der Veen, 1989. Twee modellen voor de evaluatie van de mestproblematiek. LEI Report 47. LEI, The Hague, The Netherlands.

Luesink, H.H. and A. Wisman, 2005. Mineralenverbruik uit kunstmest: in 15 jaar met 40% gedaald. Agrimonitor 11 (3): 10.

McCarl, B.A., 2006. McCarl GAMS User Guide Version 22.0. GAMS Corporation, Washington D.C., USA.

MNP, 2005. Milieucompendium 2005. Report MNP / CBS, Bilthoven, Voorburg, The Netherlands.

MNP, 2006a. Emissieregistratie. www.emissieregistratie.nl.

MNP, 2006b. Milieubalans 2006. Report MNP, Bilthoven, The Netherlands.

Oudendag, D.A., 1999. Validatie mest- en ammoniakmodel, vergelijking van de berekende ammoniakemissie bij stal- en aanwenden met metingen. Report LEI, The Hague, The Netherlands.

Oudendag, D.A. and H.H. Luesink, 1998. The manure model: manure, minerals (N, P and K), ammonia emission, heavy metals and the use of fertiliser in Dutch agriculture. Environmental pollution 102, S1, 241-246.

Oudendag, D.A. and J.H.M. Wijnands, 1989. Beperking van de ammoniakemissie uit dierlijke mest; een verkenning van mogelijkheden en kosten. Report 56. LEI, The Hague, The Netherlands.

RIVM, 1991. Nationale milieuverkenning 2 1990-2010. Report RIVM, Bilthoven, The Netherlands.

RIVM, 1995. Achtergronden bij milieubalans 95. Report RIVM, Bilthoven, The Netherlands.

RIVM/CBS, 2004. Milieucompendium 2004. RIVM/CBS, Bilthoven and Voorburg, The Netherlands.

Singh, I., L. Squire and J. Strauss, 1986. Agricultural household models: extensions, applications, and policy. The Johns Hopkins University Press, Baltimore, USA.

Smits, M.C.J., J.A. Van Jaarsveld, L.J. Mokveld, O. Vellinga, A.P. Stolk, K.W. Van der Hoek and W.A.J. Van der Pul, 2005. Het 'VELD'-project: een gedetailleerde inventarisatie van de ammoniak-emissies en -concentraties in een agrarisch gebied. Report A&F, Wageningen.

Steenvoorden, J.H.A.M., W.J. Bruins, M.M. Van Eerdt, M.W. Hoogeveen, N.J.P. Hoogervorst, J.F.M. Huijsmans, H. Leneman, H.G. Van der Meer, G.J. Monteny and F.J. De Ruiter, 1999. Monitoring van nationale ammoniakemissies uit de landbouw. Op weg naar een verbeterde rekenmethodiek. Report Staring-centrum, Wageningen, The Netherlands.

Van Bruggen, C., 2006. Dierlijke mest en mineralen 2004. Available via www.cbs.nl.

Van der Hoek, K.W., 1994. Method for calculation of ammonia emission in The Netherlands for the years 1990, 1991 and 1992 (in Dutch). RIVM Report 773004003. RIVM, Bilthoven, The Netherlands.

Van der Hoek, K.W., 2002a. Input variables for manure and ammonia data in the Environmental Balance 1999 and 2000 (in Dutch). RIVM Report 773004012. RIVM, Bilthoven, The Netherlands.

Van der Hoek, K.W., 2002b. Input variables for manure and ammonia data in the Environmental Balance 2001 and 2002 including dataset agricultural emissions 1980-2001 (in Dutch). RIVM Report 773004013. RIVM, Bilthoven, The Netherlands.

Wijnands, J.H.M. and H.H. Luesink, 1984. Een economische analyse van transport en verwerking van mestoverschotten in Nederland. Report 12. LEI, The Hague, The Netherlands.

5. Emission abatement in practical situations

André J.A. Aarnink, Hilko H. Ellen, Jan F.M. Huijsmans, Michel C.J. Smits and Dick A.J. Starmans

In The Netherlands, intensive livestock farmers are compelled to implement low emission housing systems in order to reduce ammonia emissions. In order to regulate the application of these systems, the Ministry of housing, spatial planning and the environment (VROM) has issued a list of systems that are tested and approved to be used in animal husbandry. This list is called the Rav list (VROM, 2006) and it contains emission data from traditional and low emission animal housing systems as well as emission data from housing systems that are equipped with biological and acid based air scrubber installations.

Testing of new systems that apply for the Rav list was largely performed by Wageningen UR using a standardised protocol. The protocol for these measurements, formerly known as the Green Label protocol, requires semi-continuous measurements to be carried out during two production rounds at one location. For fattening pigs this means a measurement period of about eight months. Currently, a new protocol for the admittance of new systems to the Rav list is in preparation that is expected to be brought into effect in 2007. In this protocol, special attention is given to the mode of measuring and the possibility to take into account the variance between measurement results when measuring on different locations. Chapter 6 will deal with the contents of the measurement protocol.

For the application of animal manure on grassland and arable land in The Netherlands, low emission techniques are legally prescribed. The techniques currently in use in The Netherlands are described and the corresponding emission reductions are presented in this chapter in a separate section.

5.1. Emissions from dairy cattle

5.1.1. Inventory of sources

Housing emissions

Both in numbers and emission quantity, 'dairy cattle' is by far the most important category of cattle in The Netherlands. Knowledge on emission reducing strategies in dairy cows can be ported to cattle that are kept for other purposes. Average animal emission factors for traditional housing of different categories of cattle are given in Table 5.1.

For more detailed information about specific ammonia emission measurements on cattle housing systems in The Netherlands, the reader is referred to Appendix 1. Typical traditional and low emission housing systems are described in the appendices thereafter.

In dairy cattle, traditional housing in The Netherlands refers to the most common housing system: loose housing with a slatted floor in the walking alleys and cubicles. Faeces and urine of the cows are collected in the slurry pit underneath the floor. This slurry is contributing approximately 50% of the emission from the house. Another 50% is mainly originating from fresh urine puddles on the slatted floor. The majority of beef cattle and veal calves are housed in small groups on fully slatted floors, without grazing. Nursery cows (= suckling cows) are mostly kept in deep litter (straw) housing systems and often graze during the summer season.

Table 5.1. Emission factors for some animal categories in traditional housing (VROM, 2006).

Animal category	Emission factor with grazing kg NH$_3$ (animal place)$^{-1}$yr^{-1}	Emission factor 100% housed kg NH$_3$ (animal place)$^{-1}$yr^{-1}
Dairy cattle		
Dairy cow	9.5	11
Young stock <2 yr.	3.9	3.9
Bulls for service >2 yr.		9.5
Beef cattle		
Veal calves <8 months		2.5
Nursery cow >2 yr.	5.3	5.3
Beef cattle 6-24 months		7.2

The emission factors in Table 5.1 can be used to calculate the total emission from a livestock building where a specific number of cattle per category are housed. Emission factors are also available for some other types of (adapted) animal housing to calculate total emission values for different housing situations.

The actual housing emissions depend on a number of managerial factors including:

• *Diet of the animal* (Smits *et al.*, 1995). A key factor in ammonia emission from dairy barns is the urinary urea concentration (UUC) (Elzing and Monteny, 1997; Monteny and Erisman, 1998; Monteny *et al.*, 2002). It has been shown that nutritional measures related to ammonia emission should focus on reducing UUC by reducing diet N surplus (De Boer *et al.*, 2002; Smits *et al.*, 1997). This can be realised by reducing the surplus of protein, especially rumen degradable protein, of the ration (Smits *et al.*, 1995; Smits *et al.*, 1997). The bulk milk urea content was found to be a good indicator of emission reduction (Van Duinkerken *et al.*, 2005).

• *Indoor and outdoor climate.* With increasing temperature and air velocity over emitting surfaces, emissions from these surfaces increase (Monteny, 2000). The emission from the slurry pit below a slatted floor strongly depends on the air exchange rate through the slats. This air exchange rate is especially high when relatively cool and heavy ventilation air enters the warm atmosphere in the cow house above the pit (Braam *et al.*, 1997b; Monteny, 2000). Controlling the ventilation rate, by varying inlet openings (screens or curtains) by hand or by automatic control algorithms may help to prevent high air velocities and extreme temperature differences in the cow houses.

• *Grazing regime.* During grazing hours, the ammonia emission from fresh urine on the floor of the empty cow house continues until all urea from the urine is converted to ammonia. This process takes a few hours to complete. In addition, the emission from the slurry pit continues as well, at nearly the same rate as when the cows were present. In the more extreme case where cows are kept grazing most of the day, the pit may contain a more reduced amount of ammonia that could be exhausted when continuous emissions take place.

Model calculations and emission measurements in a cow house showed an average reduction of daily cow house emissions in summer of 2.4% per daily grazing hour. When the cows are restricted in their grazing to the time during the day between milking times (approx 10 hours), staying indoors during the night, the obtained emission reduction is 24%. When the cows are grazing day and night (approx 20 hours) except around milking, average reduction of cow house emissions of 48% were calculated (Monteny *et al.*, 2001).

When kept indoors, the nutrient contents of the diet can be better balanced to the nutrient requirements of the cow than outdoors. Therefore much smaller differences between cow house emissions from grazing and non grazing herds are to be expected in practice. Yet the excrements that are collected in the cow house will also emit ammonia during and after slurry application (Smits *et al.*, 2003).

Manure storage facility
If the storage capacity in the slurry pit of the animal house is not sufficient for the whole winter season, an additional storage outside the house is needed. The emission from manure stored outside the animal house is not incorporated in the emission factors in Table 5.1. In The Netherlands liquid manure outside the house is stored in silos and manure bags. Dutch regulations have stipulated that all such structures built after 1986 must be covered.

Daily ammonia emissions from uncovered stores ranged from 2.3 g m^{-2} at 4 °C to 8.8 g m^{-2} at 25 °C (Williams and Nigro, 1997). Under Dutch conditions it can be assumed that storage facilities not equipped with a cover (built before 1987) emit at an average rate of 3.55 g m^{-2} ammonia per day during the housing season of 190 days per year. Emissions from an adequately covered storage facility are assumed to be 90% lower (De Bode, 1991; Williams and Nigro, 1997), whereas those from a rubber bag storage facility, are assumed to be 95% lower (Smits *et al.*, 2003). When the storage is used for more than 190 days per year, the emissions per day will be higher because temperatures in summer are higher.

Field emissions from grazing cattle
Bussink (Bussink, 1996) measured ammonia emission from Dutch grazed swards that were given different fertiliser rates and described ammonia volatilisation (per ha per year) in a statistical model depending on the fraction of excreted N lost as ammonia, the number of cow grazing days, and the total nitrogen excretion in faeces and urine (kg.cow^{-1}.day^{-1}). The fraction of excreted N lost through ammonia volatilisation was described in a regression function with dietary N concentration (g N/kg Dry Matter) as the independent variable (Figure 5.1).

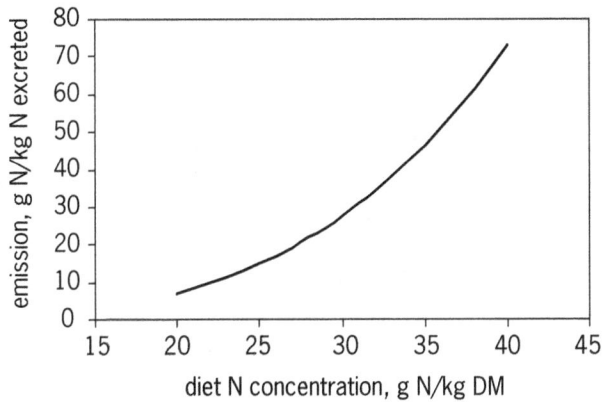

Figure 5.1. The relationship between dietary N concentration (g N/kg DM) and the fraction of excreted N lost through ammonia volatilisation (R^2=96%) during grazing on a clay soil (Bussink, 1996).

At lower fertiliser rates, the diet N concentration of grazed swards is lower, resulting in lower emissions from the grassland. By partial replacement of grass with low protein feed with a high energy content a similar effect can be obtained.

The cation exchange capacity (CEC) of the soil is influencing emission rates. On sandy soils with a lower CEC, emissions may be approximately twice the emissions given in figure 5.1. For calculations on a national scale in The Netherlands an emission factor of 8% (80 g N/kg N excreted) is used (see also Chapter 4). The number of grazing days per year and the number of grazing hours per day are decreasing in The Netherlands in the last decades. The fertiliser rate of grassland and the dietary N concentration also tend to decrease. These trends are expected to prolong in the future. Therefore ammonia emission from grazing has a small and decreasing impact on total ammonia emission.

Modelling approach

Models have been developed to describe the emission characteristics of the different sources on a dairy farm (Bussink, 1996; Huijsmans *et al.*, 2001; Monteny *et al.*, 1998). These models can be applied to quantify the variation in ammonia emission between and within farms (Smits *et al.*, 2003). Model results may be suited to set abatement priorities. Farm and environmental conditions as well as operational management are involved. Data from commercial farms were used in a study to assess the impact on ammonia emission of specific farm variables. The ammonia emissions depended on several aspects of farm management: intensity of land use, grazing management, dietary balance and slurry application have a big impact. An integrated approach is recommended in order to reduce both ammonia emission and potential nitrate leaching at farm level.

5.1.2. Emission abatement strategies

The general emission types outlined in Chapter 3 apply for keeping cattle. Theoretically, a number of ammonia abatement strategies are available in this field. The most relevant principles studied to reduce cow house emissions in The Netherlands are:
- Reduction of emitting surface. Exposed manure and urine are a distinct emission source. Reduction of one or both sources leads to lower emissions to the environment.
- Adaptation of airflow along manure surfaces. By reducing the airflow along manure surfaces, the mass transfer of ammonia from the manure liquid to the gaseous boundary layer and from this layer to the ventilated air is decreased.
- Feeding management strategies to manipulate excretion characteristics from animals inside the animal house.

In the past also some other principles were tested in experiments in dairy cow houses. Especially slurry handling systems like flushing with water (reducing urea concentration) and with a formaldehyde solution (reducing urease activity) were tested (Ogink and Kroodsma, 1996). Also acidification of slurry in the pit with nitric acid, reducing pH of slurry < 4.5 to prevent denitrification, and additionally flushing the slats with acidified slurry (also reducing pH) was tested (Bleijenberg et al., 1995). However, these systems were never adopted on commercial dairy farms because of practical disadvantages like increasing the volume of slurry (requiring additional storage capacity and extra costs for slurry application), health risks with formaldehyde and N surpluses due to adding nitric acid. Other organic or inorganic acids may be considered as an alternative for nitric acid. However, it is essential that care should be taken when working with acids at the farm. Administration rates should be well controlled and mixing should be done safe and properly as to not only to obtain the best results but also to prevent peak levels of toxic gases like hydrogen sulphide and cyanic acid gas. A more detailed overview of possibilities for reduction that were studied in the past is given in literature (Monteny and Erisman, 1998).

5.1.3. Practical solutions

Solid floors

The ammonia emission reduction of several solid floor designs was studied in detail (Braam *et al.*, 1997a,b; Swierstra *et al.*, 2001). The air velocity above the surface of slurry in the pit and the air exchange between the pit and the house play a key role in the ammonia emission from the pit. Both can be reduced by covering the slurry pit with a solid floor. In the case of perfect covering of the slurry pit, its contribution to ammonia emission from the building may be eliminated. In case the manure is dropped in an under floor pit, the manure collected by a scraper is dropped into the pit through openings at the floor ends and sometimes halfway. Due to the necessary floor openings, the emission reduction is lower than in case of completely air tight covering of the pit. The design of these openings is crucial for the air exchange and ammonia emission. In experiments emission reductions above 50% were found compared to a reference cubicle cow house unit with a slatted floor (Swierstra *et al.*, 1995). In commercial farms emission levels of cow houses with V-shaped solid floors were only approx 20-25% lower than the emission level of a traditional cow house. In Appendix 3 some more details of these floors are given.

After solid sloping floors were introduced on commercial farms, welfare problems related to the slipperiness were reported especially in summer when the cows came in after grazing. By grooving the floor this problem can be tackled, however, the emission reduction than will be less (Smits, 1998).

Another solution with precast concrete floors with grooves and a dung scraper was investigated (Swierstra *et al.*, 2001). The grooves parallel to the alley had 160 mm centre-to-centre spacing and were 35 mm wide and 30 mm deep. Urine could drain directly into a slurry pit below through small perforations in the grooves that were spaced 1.1 m apart. The faeces were dragged to the end of the alley using a scraper, provided with facilities that also cleaned the grooves. The faeces were dropped into the pit through an opening at the floor. This opening was specially designed, like a mailbox with a lid, to minimise air exchange. This system is described in more detail in Appendix 4. The system optionally can provide primary separation of urine and faecal matter. Since urine is a major source of nitrogen, this approach may lead to less emission of ammonia. Furthermore the separated fractions can be utilised more efficiently.

Feeding management
Smits *et al.* (1995) showed the potential for reducing ammonia emissions
by reducing the rumen degraded protein surplus in animals kept in a forced
ventilated building. Besides the protein surplus in the di*et al.*so the type
of roughage may play a part in ammonia emission. It may have an effect
on the nitrogen intake level, its utilisation and excretion in faeces and
urine. Also it may influence the concentration of excreted minerals due
to surpluses of potassium and sodium in the diet that mainly determine
the urine volume (Bannink *et al.*, 1999; Van Vuuren and Smits, 1997). On
commercial farms in The Netherlands grass and maize are the main types of
roughage that are fed in varying proportions depending on the region and
local farm management. The effects of both rumen degraded protein and
type of roughage were studied in a naturally ventilated cow house. Results
are summarised in Figure 5.2.

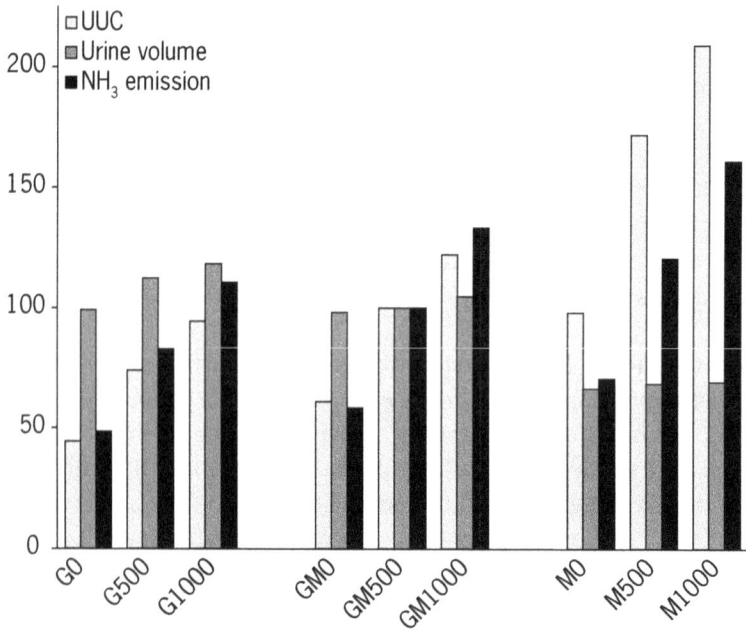

*Figure 5.2. Proportional effects of roughage type (G=grass silage, M= maize silage,
GM= 50% grass silage, 50% maize silage) and rumen degraded protein surplus (0,
500 or 1000 g per day) on urinary urea concentration (UUC), urine volume and
ammonia emission in a two year experiment where diets were alternately fed.
Outcomes were expressed as percentages of those with diet GM500. Data from Van
Duinkerken* et al. *(2005).*

For both farmers and government a tool for monitoring the level of ammonia emission would be useful. From the known systemic pathways of urea it was hypothesised that bulk milk urea could be such an indicatory tool. Its relation with ammonia emission was studied. Results are summarised in Figure 5.3.

As stated in Chapter 2, the implementation of the IPPC guidelines resulted in the mandatory introduction of low emission housing systems for Dutch poultry and pig farmers. In 2002, the Dutch government came to an agreement with the dairy farmers in The Netherlands: in exchange for farmers willing to reduce milk urea content to an average value of 20 mg urea per 100 ml milk in 2010 by improving the dietary balance of protein, they were safeguarded against having to invest in low emission housing systems until 2010. The agreement was made after preliminary results indicated that there were good prospects for an approximate 25% reduction of ammonia emission by feeding management on commercial farms (Van Duinkerken *et al.*, 2003; Van Duinkerken *et al.*, 2005).

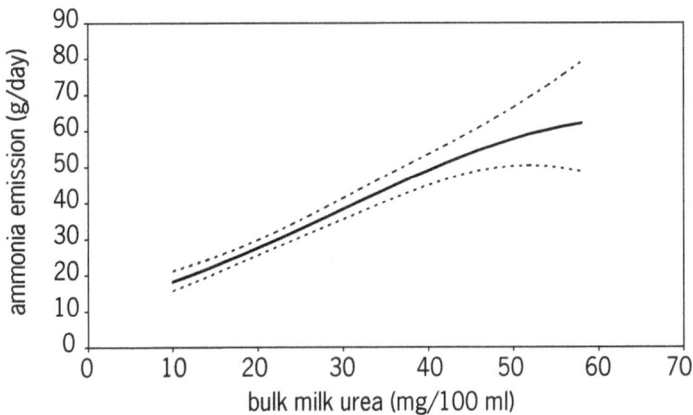

Figure 5.3. Relationship between bulk milk urea content (mg/100 ml) and ammonia emission. Ammonia emission is corrected to an outside temperature of 15 °C (approx 2.7% per °C). With dotted lines 95% confidence limits, based on the accuracy of the model parameters estimated within the data of this experiment, are indicated (Van Duinkerken et al.*, 2005).*

5.1.4. Variations in ammonia emissions

As compared to pigs and poultry, cow housing systems display a larger variation of emissions due to natural ventilation, nutrition and other management aspects. An inventory of the ammonia emission of ammonia in several livestock buildings in four North European countries: England, The Netherlands, Denmark and Germany was made (Groot Koerkamp *et al.*, 1998). Emissions from the most relevant housing systems were determined with a limited number of short term measurements (approximately 4 days per season) in summer and winter on four commercial farms in each country. Ammonia emission results with regard to cattle of this EU project are given in Table 5.2.

Large variations within and between experimental units were found. These variations were not only due to systematic effects that differ between countries or between housing systems but these variations were also caused to some extent by chance (e.g. meteorological conditions during short measurement period, diets fed, animal production level).

Table 5.2. Ammonia emission from different housing systems in a quick scan (Groot Koerkamp et al., 1998). Emission of ammonia (mg/h per animal) for the mean outside temperature, coefficient of variation (%) for various housing systems for cattle in England (10.1 °C), The Netherlands (9.8 °C), Denmark (8.4 °C) and Germany (10.5 °C).

Cattle type, housing system	England		The Netherlands		Denmark		Germany	
	mean	c.v. %	mean	c.v. %	mean	c.v. %	mean	c.v. %
Dairy cows, litter	314[a]	45	974	24	560	24	538[a]	31
Dairy cows, cubicles	1245[a]	52	2,001	24	987[a]	25	1,320	31
Beef cattle, litter	482[a]	48	na		na		262	27
Beef cattle, slats	na[b]		686	24	580	22	346[a]	31
Calves, litter	80[a]	41	na		332	23	193	24
Calves, slats/group	na		522	24	na		323	24

[a]Measuring inaccuracies because outside NH_3 concentration >= 20% of inside NH_3 concentration.
[b]na: not available, mostly because system is not common in The Netherlands.

5.2. Emissions from pigs

5.2.1. Inventory of sources

Pig emissions

Average animal emission factors for traditional housing are given in Table 5.3. When pigs are kept in a more confined space, the emission factor is generally decreasing.

These emission factors can be used to calculate the total emission from a livestock building where a specific number of animals from each category are housed. They therefore include emissions from the pen (i.e. soiled surface area) and pit. Factors are also available for several types of (low emission) animal housing to calculate total emission values for different housing situations.

The actual housing emissions depend on a number of managerial factors including:
- *Diet of the animal.* Nutrition is a key factor to reduce environmental load from pig production (Aarnink and Verstegen, 2007). Nutrition mainly influences ammonia emission by influencing the ammonia content and the pH of the manure.

Table 5.3. Emission factors for some animal categories in traditional housing (VROM, 2006).

	Emission factor, kg NH_3 (animal place)$^{-1}$yr^{-1}	
Animal category	**Large pen**	**Small pen**
Weaned piglet < 25 kg[1]	0.75	0.60
Fattening pigs[2]	3.5	2.5
Farrowing sows (incl. piglets)	8.3	
Dry and pregnant sows	4.2	
Boars >7 months	5.5	

[1]Large pen: >0.35 m²/animal, small pen: <0.35 m²/animal.
[2]Large pen: >0.8 m²/animal, small pen: <0.8 m²/animal.

• *Indoor climate.* In pig houses the indoor climate is mainly temperature controlled. Depending on the outside temperature the house is more or less ventilated and more or less heated, to keep the inside temperature within certain limits. Research showed a clear influence of temperature and ventilation rate on ammonia emissions (Aarnink *et al.*, 1996; Häußermann *et al.*, 2005). In field studies these two variables are almost always intertwined with each other and with animal weight. However, model studies showed clear effects of these variables on ammonia emission from pig houses (Aarnink and Elzing, 1998).

• *Application of bedding material.* In a few studies the effects of bedding material on ammonia emission have been studied. Although the use of a limiting amount of straw or other bedding material might have a negative effect on ammonia emission, a big layer of straw in the lying area can have a positive effect (Groenestein, 2006).

Manure storage facility

In The Netherlands, research has lead to the general rule that liquid manure has to be stored in closed facilities (Williams and Nigro, 1997). Furthermore, farmers are obligated to have silo's or manure bags on their farms for temporal storage of manure. In general, the emissions from these facilities are much lower than the emissions from the animal house. It is therefore that abatement strategies predominantly focus on the latter.

5.2.2. Emission abatement strategies

As seen in Chapter 3 a number of emission types can be discerned. In the case of pig production, a number of emission abatement strategies are available to counteract emissions of ammonia from the animal house to the environment:

• *Reducing the fouled floor area.* Ammonia emission is linearly related to the area fouled with urine and faeces.

• *Reduction of manure surface.* The emission from a liquid manure surface is proportional to its surface area.

• *Nutritional measures.* Ammonia emission is linearly related to the ammonium concentration in the manure and exponentially related to the pH of the manure. By nutrition, both, the ammonium concentration and the pH of the manure can be influenced.

• *Manure management.* Regularly flushing or scraping the manure from the manure pit can reduce ammonia emission. Lowering the temperature of the manure or diluting the manure with water or with aerated liquid manure also reduces ammonia emission. The liquid manure can also be treated with formaldehyde to bind ammonia, and use this liquid for flushing.

• *Avoiding ammonia forming reactions.* The enzyme urease in solid manure accelerates the break down of urea (present in urine) into ammonia and carbon dioxide.

5.2.3. Practical solutions

Reducing the fouled floor area

A well designed pen and a good indoor climate is essential to reduce the fouled floor area in pig housing. A lot of work has been done in this area during the last decades. New pen designs have been developed to reduce fouling of the floor (Aarnink *et al.*, 1996; Hacker *et al.*, 1994; Hol and Satter, 1998; Reitsma and Groenestein, 1995). An important factor with respect to pen fouling is the inside temperature (Aarnink *et al.*, 2006). By using cooling systems pen fouling during the summer can be prevented (Huynh *et al.*, 2004).

Reduction of manure surface

Reduction of manure surface area can be decreased by changing the geometry of the manure pit. Newly built animal housings are often equipped with a modified pit. While in traditional housing the pit is placed under the complete pen, in new housings the maximum surface area reserved for the pit is often reduced to 60% or less. Application of inclined pit walls can further reduce the emitting manure surface area. An example of a system with reduced pit area and inclined pit walls is given in Figure 5.4.

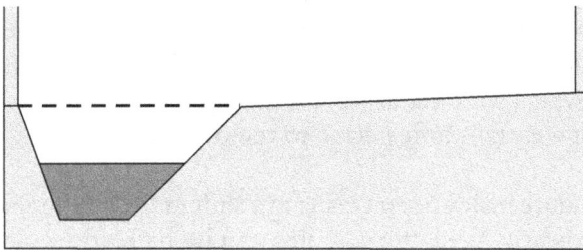

Figure 5.4. Cross view of a pen with modified pit geometry.

Nutritional measures
Main nutritional strategies to reduce ammonia emissions are: (1) lowering crude protein intake in combination with supplementation of limiting amino acids (Canh *et al.*, 1998b); (2) shifting nitrogen excretion from urine to faeces and simultaneously lowering the pH of manure by including fermentable carbohydrates in the diet (Canh *et al.*, 1998c,d, 1999; Sutton *et al.*, 1999); (3) lowering pH of urine by adding acidifying salts to the diet (Canh *et al.*, 1998a). These strategies proved to be independent from each other and effects are additive. By combining these strategies a total reduction of ammonia emission in growing-finishing pigs of 70% could be reached (Bakker and Smits, 2002).

Manure management
Regular flushing of manure from pit to storage facility leads to a decreased manure surface area exposed to ventilated air. Flushing can be realised by pumping water or acidified liquid manure. It can be stimulated using gutter-like pit configurations with drains on the lowest points.

Another way of reducing ammonia concentration is by taking up the voided manure into a liquid such as water or formaldehyde treated liquid manure. In the former, the ammonia is only diluted, whereas in the latter, the ammonia and urea can react with the formaldehyde according to Equation 1.

$$R\text{-}NH_2 + H_2CO \rightleftharpoons R\text{-}NH\text{-}CH_2OH \tag{1}$$

With group R being H in the case of ammonia or $H_2N\text{-}CO\text{-}$ in the case of urea.

Primary separation of solid and liquid manure can be used to counteract enzymatic ammonia generation. The most commonly used method of separating solid manure from urine is the use of tilted or convex conveyor belt systems, which continuously allow the liquids to be drained off. The belt is moved after a certain time period to remove the gathered solid manure.

Cooling of manure inside the pit leads to a shift in the ammonia-ammonium equilibrium towards the latter. Cooling can be realised by placing cooling surfaces (heat exchangers) inside the pit. The warmed-up cooling liquid is cooled elsewhere, before it is pumped to the heat-exchanger again.

Housing systems

A general overview of the change in emission factors by the application of each of these techniques is given in Table 5.4. The systems listed are available through third party companies and are taken up in the Rav list.

As can be seen in Table 5.4, the reduction obtained by chemically modifying the manure ammonia is similar to that obtained by using a strongly adapted pit geometry and subsequent decrease in manure surface area. Pig farmers tend to opt for the latter since it is an inherent system that is not requiring much attention or additional chemical knowledge.

Table 5.4. Emission factors for different housing systems (data for fattening pigs) (VROM, 2006).

System	Description	Emission factor[1] (kg NH_3.animal place^{-1}.yr^{-1})
Traditional	Fully slatted floor	4.0
Flushing	Flushing with NH_3 poor or acidified fluids	1.0 – 2.0
Cooling	Cooldeck system 170% cooling surface	1.4 – 2.0
Chemical	Manure take-up in formaldehyde treated liquid manure	0.8 – 1.1
Dilution	Manure take-up in water	1.1 – 1.5
Cooling	Cooldeck system 200% cooling surface	1.0 – 1.4
Pit geometry	Inclined pit wall, max. surface 0.18 m^2.animal^{-1}	1.0 – 1.5

[1]Depending on square footage per animal and slatting material.

5.3. Emissions from poultry

5.3.1. Inventory of sources

Poultry emissions

Average poultry emission factors for traditional housing are given in Table 5.5. The wet belt systems were applied to get some basic ammonia emission abatement. In these systems, manure is transported via belts to a confined space at least two times a week. The average ammonia emission from poultry kept in housing with these systems is reduced by more than 50% in comparison with open manure storage under the cages.

Table 5.5. Poultry emission factors in traditional housings in kg NH$_3$ (animal place)$^{-1}$yr^{-1} (VROM, 2006).

Animal category	Cages deeppit or highrise	Cages open manure storage	Cages wet belt system	Deep litter
Layers < 18 weeks	0.208	0.045	0.020	0.170
Layers + layer breeders[1]	0.386	0.083	0.035	0.315
Broiler breeders < 19 weeks				0.250
Broiler breeders				0.580
Broilers				0.080
Turkeys				0.680
Ducks; outside keeping				0.019
Ducks; inside keeping				0.210

[1]Values mentioned are the old values. Because of the demand for extra living area, these values have been increased by 20% to respectively 0.463, 0.100 and 0.042 kg NH$_3$ (animal place)$^{-1}$yr^{-1}.

Feed rations

Feed rations are normally stored outside the animal house. Poultry is kept on dry rations that are expected to have low emission values due to the low protein content. Further decrease in emission by decreasing the protein content even more is not an option since this would cause a decrease in animal performance as well (Ellen *et al.*, 2005).

Manure storage facility

If poultry is kept in a deep litter system the solid manure stays in the house for the whole production period. If manure belts are used (as in cage or aviary housing) the manure is stored outside the house in closed containers. This manure is stored for an average period of two weeks, after which it is transported. Emission factors for these systems are also available in the Rav list (VROM, 2006).

5.3.2. Emission abatement strategies

In the case of poultry keeping, a number of emission abatement strategies are available to counteract the emission of ammonia from the animal house to the environment:

- *Manure management.* By regularly removing the excrements to a closed storage area, the net ammonia emission to the environment is strongly reduced. Belt systems can be used to transport manure to such a centralised storage area.
- *Drying of poultry manure.* Further emission abatement techniques proved to be necessary because of the increasing environmental load with increasing poultry population. The drying of poultry manure proved an effective way to reduce the ammonia emission (Kroodsma et al., 1985).
- *Use of litter.* In free range systems, litter is often used for taking up the excrements. The physical and chemical properties of litter influence the ammonia emission. By either choosing a certain type or modifying an existing litter, the ammonia abatement properties can be enhanced to suit the needs (Huff, 1984).

5.3.3. Practical solutions

Housing geometry
The manure surface area in free range systems is rather large and only allows for the use of litter. The other end of the spectrum is formed by cage systems, where the manure surface area is limited to a small belt under the cages. Taking the beneficial things of both these systems, the aviary system provides enough space for natural poultry behaviour and includes an efficient manure removal system similar to that observed in cage systems.

Manure drying
On-site drying of poultry manure can be realised in many ways. In most systems, the manure is dried on belts using air that is warmed-up by the animals present. Litter systems can also be equipped with a drying system. The distribution of warmed air over the relatively large area is realised using distributing hoses (Figure 5.5). The emission factors of these systems are listed in Table 5.6.

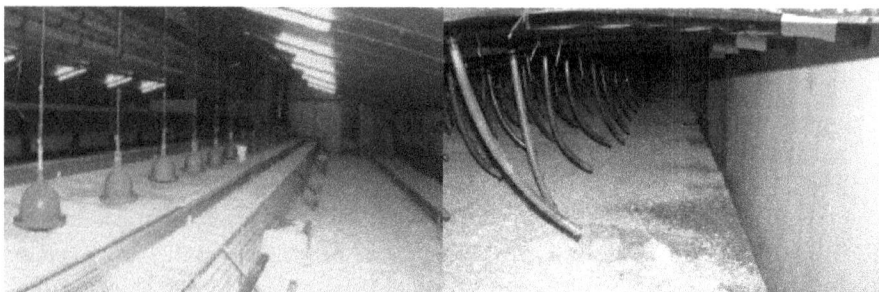

Figure 5.5. Free range litter system (left) with drying of litter underneath the feeding and drinking area of a broiler breeder house (right).

Table 5.6. Adapted systems and related emission factors for layers, layer breeders, broiler breeders and broilers with forced manure drying (VROM, 2006).

System	Emission factor (kg NH_3 (animal place)$^{-1}$yr^{-1})
Layers and layer breeders	
Battery system with forced air drying	
0.5 m^3/hen/hour	0.042
0.7 m^3/hen/hour	0.012
Deep litter system with air drying through hoses	0.125
Deep litter system with slatted floor underneath the manure	0.110
Aviary systems	
with 50% slatted floor, 0.2 m^3/hen/hour	0.055
with 30-35% slatted floor, 0.7 m^3/hen/hour	0.025
with 55-60% slatted floor, 0.7 m^3/hen/hour	0.037
Broiler breeders	
Deep litter system with air drying through vertical hoses	0.435
Deep litter system with air drying through horizontal hoses	0.250
Deep litter system with slatted floor underneath the manure	0.230
Broilers	
Raised floor with litter drying	0.005
Perforated concrete floor with litter drying	0.014
Turkeys	
Partly raised floor with litter drying	0.360

5.4. Emissions after manure application

5.4.1. Inventory of sources

The ammonia volatilisation rate of nitrogen applied with animal manure is a function of many parameters. Only the total ammoniacal nitrogen (TAN) in animal manure is subject to volatilisation. Nitrogen in liquid manure contains normally 45-55% TAN, whereas nitrogen in solid manure contains 10-40% TAN. The total ammonia emission depends on the total amount of TAN applied, the application method, the manure characteristics, and on the field and weather conditions (Huijsmans, 2003). With respect to the weather conditions volatilisation increases when radiation, wind speed, and air temperature are higher. On the other hand rainfall has a reducing effect on the ammonia volatilisation.

Nowadays the total Dutch animal manure production of 70 million tons per year is divided into roughly 50 million tons of liquid manure and 4 million tons of solid manure, supplemented by 15 million tons manure deposited directly in the meadow by grazing cattle (see Chapter 1, Table 1.1).

5.4.2. Emission abatement strategies

Key parameters in the abatement of ammonia emissions are the decrease in the emitting surface area and the reduction of the contact time between animal manure and ambient air. For this purpose, several different application techniques have been developed over the years. In The Netherlands, narrow band application by trailing feet or shoes and shallow injection were made compulsory in the 1990s for grassland. The same was the case for direct incorporation of surface-applied manure and injection of manure into arable land. Starting in 1990 with regulations for sandy soils, from 1995 broadcast surface spreading was no longer allowed in The Netherlands. The share of the different low emission techniques nowadays in use in The Netherlands is presented in Chapter 4 (see Table 4.4).

Starting in the same period of 1990-1995, manure application in autumn and winter was no longer allowed. Nitrogen applied in these seasons is subject to leaching and therefore not available for the growing crop in the next spring. This partly reduces the effect of the emission reducing techniques, since conditions favouring ammonia emissions (less rainfall, more radiation, higher temperatures) are more often met in spring and summer than in autumn and winter. Therefore, when comparing the overall national annual ammonia emissions between the 1980s and the period from 1990 onwards, not only the application methods used, but also the time of the year when manure was applied should be taken into account (Huijsmans *et al.*, 2001; Huijsmans *et al.*, 2003).

5.4.3. Practical solutions

Broadcast surface spreading of manure on grassland and arable land used to be the common method for manure application. It was carried out by a tanker fitted with a splash-plate. The liquid manure was pumped through an orifice onto a splash-plate from where it was spread onto the soil and the grass.

Grassland
Techniques for the application of liquid manure on grassland were described in literature (Huijsmans *et al.*, 1998). The technique of deep injection on grassland is no longer in use in The Netherlands, it is succeeded by shallow injection (open slot).

Narrow band application is carried out by trailing narrow sliding feet (also called 'shoes') over the soil surface, pushing aside the grass cover but not cutting the sward. Manure is released at the back of the feet leaving narrow bands of manure onto the soil surface. The bands had a width of about 0.03 m and are spaced 0.20 m apart.

Shallow injection (open slot) is carried out with injection coulters. Coulters and discs are used to cut vertical slots into the grass sward. Manure is released into the slots, which were left open. The slots are up to 0.05 m deep and are spaced 0.20 m apart. Depending on the application rate, the slots are more or less filled with manure. Unlike the conventional deep injector, the shallow injectors have no lateral wings and do not cut the soil horizontally underneath the sward (see Figure 5.6).

Arable land
Techniques for the application and incorporation of liquid manure on arable land are described in literature (Huijsmans *et al.*, 2003).

Surface incorporation is defined as the treatment in which manure is surface applied, followed by incorporation into the soil. Conventional tillage

Figure 5.6. Shallow injection of liquid animal manure on grassland. The left picture shows the shallow injector with the circular discs for making a vertical slot in the soil and the openings for manure release. The vertical slots are up to 0.05 m deep and are spaced 0.20 m apart. The right picture shows the field, injected at left and yet untreated at right. Note that the slots remain open. (Photos by courtesy of Erik van Asten, Leende)

implements (cultivators with rigid tines, spring tines, discs, or harrows) are used to incorporate the surface applied manure into the topsoil directly after surface spreading.

Deep placement is defined as the treatment in which the manure is buried in the soil, either directly by an injector or indirectly by ploughing with a mouldbourd plough directly after surface spreading. The arable land injector is equipped with spring tines, which place the manure directly underneath the soil surface at a depth of 15 to 20 cm. At the same time the injector carries out a tilling operation covering the manure with soil.

5.4.4. Practical aspects of field measurements

To measure emissions during the application of liquid manure, manure was applied on a small circular area and the emission was measured with two masts, one in the middle of the field and the other outside the field (Figure 5.7).

Most experiments were comparative, which means that surface application of animal manure was compared to low emission techniques such as deep injection, shallow injection or direct incorporation. An advantage of this approach is the easy elimination of weather, soil, and manure type differences during the experiment. A wide range in emission reductions was obtained,

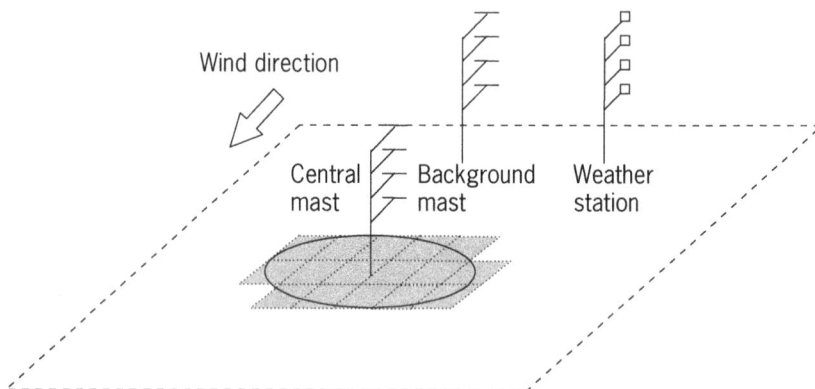

Figure 5.7. Lay out of circular plot (diameter about 50 m) for the measurement of ammonia volatilisation using the micrometeorological mass balance method, with masts supporting NH₃ traps at various heights in the centre of the plot and at the windward boundary of the plot (Huijsmans et al., 2001).

partly caused by the prevailing weather conditions during the individual experiments. Also the period during which the animal manure is exposed to the atmosphere is very important as is the case with incorporation in arable land. In practice on whole field scale, direct incorporation is not always achievable. There will always be some time between surface spreading and incorporation. During this time volatilisation of ammonia from the surface-applied manure takes place (Huijsmans and De Mol, 1999). They showed that the time lag between spreading and incorporation should be considered when assessing ammonia volatilisation from manure applied and incorporated on arable land. In case of deep placement by injection the time lag is zero and low volatilisation rates can be achieved.

The average ammonia emissions on grassland and arable land are presented in Table 5.7, together with the observed band width (state of art 1999). In Dutch national emissions inventories these average emissions were multiplied by a factor 1.15, because under practical conditions emissions were expected to be higher due to a difference of scale. The cumulative ammonia losses after manure application can be calculated according to the application method and the type of land use.

Fundamental chemical and physical processes were used as a starting point for modelling ammonia volatilisation from arable land after application of

Table 5.7. Cumulative ammonia losses after manure application depending on the method and the land use type, expressed as a percentage of the TAN applied.

Application method	Land use	% of TAN lost[1]	Band width[1]	% of TAN lost[2]
Broadcast surface spreading[3]	grassland	68	27 - 98	68
Narrow band application	grassland	25	8 - 50	28.75
Shallow injection	grassland	10	1 - 25	11.5
Broadcast surface spreading[3]	arable land	68	20 - 100	68
Surface spreading and incorporation in two tracks	arable land	20	1 - 49	46
Surface spreading and incorporation in one track	arable land			23
Injection	arable land	9	0 - 40	10.35

[1]Data from (Huijsmans, 1999).
[2]Data from (Van der Hoek, 2002), used in Dutch emission inventory calculations. Data are based on Steenvoorden *et al.*, 1999 and on the minutes from a meeting afterwards between specialists on animal manure application.
[3]In the period before 1990 calculations used 50% in stead of 68% TAN loss. After 1990 animal manure is only applied in spring and summer during which weather conditions favour higher ammonia emissions.

cattle manure (Van der Molen *et al.*, 1990a,b). The model however required the input of many detailed parameters that could hardly be collected at farm scale.

All data from the aforementioned field experiments (Table 5.7) were further statistically analysed to assess a better estimation of emissions after manure application by different manure application techniques, and to assess the main factors and conditions that affect the ammonia volatilisation following manure application. Besides the application method and the land use type (grassland or arable land), the impact of conditions such as field and weather conditions, and manure characteristics were taken into account (Huijsmans, 2003). In any situation ammonia volatilisation from animal manure applied on farmland is substantially reduced by appropriate techniques for the application and incorporation of manure. Actual environmental conditions under which manure is applied, including field and weather conditions, manure composition and manure application rates, also substantially affect the overall ammonia volatilisation. Thus, when reliable predictions of ammonia volatilisation are required, for example for farm management or for a national approach to abate ammonia volatilisation, both the techniques for application and incorporation, and factors influencing ammonia volatilisation must be taken into account (Huijsmans *et al.*, 2001, 2003).

Presently, an actualisation of ammonia volatilisation factors for the different manure application techniques is part of a study in which also the more recent ammonia volatilisation measurements are taken into account.

5.5. General end of pipe abatement techniques

5.5.1. Inventory of sources

Agricultural production includes the production of waste and ammonia emissions. The previous Sections 5.1, 5.2 and 5.3 focussed on emission reduction related to animal management and housing. All remaining emissions can be seen as sources for the additional end of pipe abatement techniques described here. However, all end of pipe technical solutions to reduce ammonia emissions require mechanically ventilated housing systems with well defined exhaust air vents.

A general overview of the change in emission factors by the application of end of pipe techniques is given in Table 5.8. The systems listed are available through third party companies and are taken up in the Rav list.

Table 5.8. Emission factors for different end of pipe techniques (data for fattening pigs) (VROM, 2006).

System	Description	Emission factor[1] (kg NH_3.animal place^{-1}.yr^{-1})
Traditional	fully slatted floor	4.0
Biological scrubber	take-up of ammonia into water followed by bacterial breakdown to nitrate and nitrogen	0.8 – 1.1
Chemical scrubber 70%	take-up of ammonia into acidified water	0.8 – 1.1
Chemical scrubber 95%	take-up of ammonia into acidified water	0.13 – 0.18
Combined chemical and biological	take-up of ammonia into acidified water and subsequently placed biofilter unit	0.75 – 1.05

[1]Depending on square footage per animal and slatting material.

5.5.2. Emission abatement strategies

The most common end of pipe ammonia abatement technique is the application of an air scrubber system for the removal of ammonia from animal houses. These systems are analysed for their performance in literature (Melse and Willers, 2004; Scholtens, 1996). Both biological and acid based scrubbers can be utilised for the removal of ammonia from exhaust air from animal houses. Typical ammonia abatement rates of these systems are <70% and >95% respectively. This means that the latter are tested using the *in-situ* protocol explained in Chapter 6, whereas the former systems are evaluated theoretically.

The use of wet techniques to reduce ammonia emissions also has other beneficial aspects. Air passing a chemical or biological scrubber is also purified with respect to dust and pathogens (Aarnink *et al.*, 2005).

Current developments in this field of end of pipe techniques include combination air scrubbers, in which a chemical washing section is preceding a biological washing section. In this way both ammonia and odorous compounds can be eliminated from the treated air.

5.5.3. Practical solutions

Acid scrubber
Although gaseous ammonia is soluble in water (27.8 mol/l at 25 °C), it is nearly completely soluble in acidified water (base constant of ammonia = $1.8. 10^{-5}$ mol/l, which accounts for a maximum of 99.6% of the ammonia to be in the protonated form when the water phase is sufficiently acidified). For this reason, acid scrubbers are used for the purification of exhaust air.

Biological scrubber
Though maybe not as efficient as acid scrubbers, biological scrubbers can also be used as a means of capturing ammonia from exhaust air from livestock housing. Without additional carbon source, some immobilised, growing biomass can incorporate ammonia. Surplus biomass is purged from the system.

Key parameters required for the theoretical evaluation of acid or biological scrubbers are given in Table 5.9. A general layout of scrubber systems is given in Figure 5.8.

Table 5.9. Key parameters for the theoretical evaluation of scrubbers.

Functional description
- intended ammonia removal (%)
- type of exhaust air? (pigs, poultry)
- flow: counter, co or cross-current?
- maximum air flow (m^3/h)
- water recirculation flow (m^3/h)
- purge flow (m^3/h)

Scrubber specific design
- type of spray nozzles / specification
- packing material: brand name, type, specific surface (m^2/m^3), HTU value (m), pressure drop, etc.
- volume of packed section (m^3)
- cross sectional surface (m^2)
- height (m)
- volume of buffer tank

Specific info for biological scrubbers
If a separate biological section exists, separate from the scrubbing section:
- maximum NH_3 removal capacity of system (g m^{-2} hr^{-1})
- specific surface of packing material of biological section(s) (m^2/m^3)
- volume of biological section(s) (m^3)
- water flow through biological section (m^3/uur)
- temperature range of operation (winter, summer, measures to be taken)

Specific info for acid scrubbers
- type of acid used
- description of acid dosing system
- description of pH and EC measurements
- frost prevention on acid supply

Dust and suspended materials
- clogging precautions / dust removal before the air enters the scrubber
- packing material cleaning procedure
- solids removal from the buffer tank

Miscellaneous
- detailed description of all control systems that are used for operation

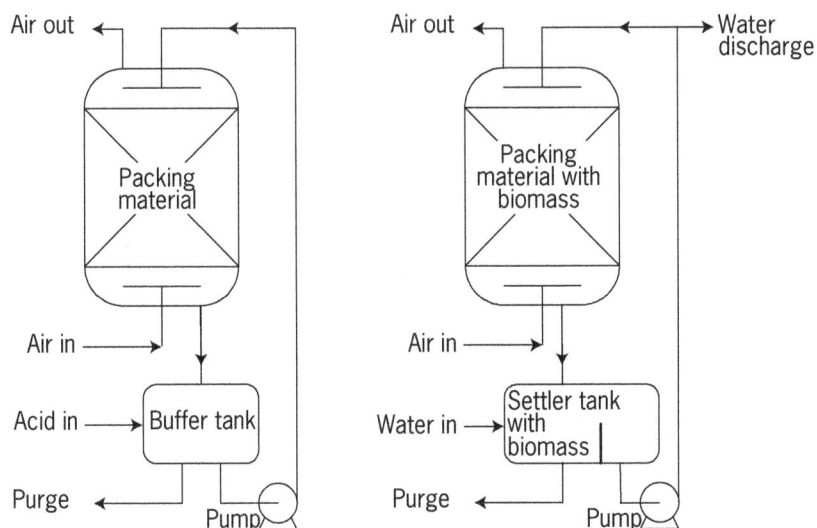

Figure 5.8. General layouts of an acid scrubber (left) and a biological scrubber for the removal of ammonia from exhaust air (right).

In the specific case of poultry housing systems, end of pipe ammonia emission abatement is somewhat modified with respect to the normal operation using belt systems to transport the manure to storage facilities outside. These extra measures to further decrease the ammonia emission from stored solid manure include:
• drying the manure inside a heated tunnel;
• composting the manure in a closed unit;
• application of additional drying air.

The remaining emissions from the stored solid manure are added to the housing system emission factor. Using these techniques, the emission from the stored solid manure could be reduced from 50 grams per animal per year to several grams of ammonia.

5.5.4. Theoretical evaluation

The theoretical evaluation is based on theoretical and experimental derived relations regarding mass transfer in columns. The main part is an estimation of the Height of Transfer Unit (HTU) based on the overall mass transfer from the gas to the liquid phase. The Number of Transfer Units (NTU) is calculated from the desired removal percentage. For 70% removal, NTU is 1.2 and for 95%, NTU is 3. The minimum packet size or column height necessary to obtain the desired removal percentage, is calculated as HTU x NTU. Column volume is modified for incomplete wetting and / or flooding of the packing material by means of estimation. The calculations mentioned above are described in engineering handbooks (Sinnot, 1999).

For many common random packing materials, e.g. Pall rings, Raschig rings, etc., characteristics have been experimentally determined in the past. These characteristics include packing factor (m^2/m^3), sphericity (-), specific surface area (m^2/m^3), and bed porosity (-). Based on these characteristics the HTU value can be calculated as described above.

However, many companies design scrubbers that use structured packing materials as structured packing materials in general have a lower pressure drop and a higher mass transfer efficiency, which means a lower HTU. Unfortunately, the characteristics of many of these packing materials are not well known. In these cases, it depends on the amount and quality of the additional information that is given by the applicant, e.g. lab experiences, measurements, etc., if a theoretical estimation can be made.

To correct for non-ideal circumstances and variations in airflow and ammonia concentration, the estimated necessary column height is multiplied with 1.5 to 2 to be sure that the intended ammonia removal will also be achieved in a real life situation. With regard to clogging and dust removal, it is stated that the company that sells the scrubber is responsible for taking sufficient measures to prevent accumulation of solids in the system. In our opinion, if an air scrubber is designed well, it is necessary to clean its packing at most twice a year.

Ammonia load
The ammonia load of a scrubber is assumed to be equal to the ammonia emission figures that are stated in the Rav list for housing systems without air scrubbing (VROM, 2006). Some examples from this list are mentioned throughout this chapter in Tables 5.1 and 5.3 till 5.6.

Biological scrubber
In case of a biological scrubber, the ammonia transfer from gas to liquid phase is followed by microbial oxidation to nitrite and nitrate. Because of the equilibriums of these processes, it is assumed that the maximum ammonia

removal capacity is 70% at reasonable water discharge rates. If a biological scrubber system is claimed to achieve higher removal percentages than 70%, a theoretical evaluation is not sufficient and additional measurements are required.

The maximum ammonia removal capacity of the biological scrubber system is calculated based on an estimation of the biomass concentration in the water phase, the nitrification rate of the biomass, the liquid hold up in the scrubbing section, the volume of the recirculation settler tank, and the water discharge rate. This capacity should be sufficient to remove 70% of the ammonia load. The design of the scrubber together with the water discharge rate must prevent accumulation of nitrite, nitrate, and ammonia which is necessary to assure a stable operation.

Acid scrubber
In an acid scrubber the ammonia is ionised by the acid after being transferred from the gas to the liquid phase. In a well designed scrubber operating at low pH, ammonia removal percentages of over 95% can be achieved. Currently, as for biological scrubbers, a theoretical evaluation for an acid scrubber design is only used for approval if the claimed removal percentage is lower than or equal to 70%.

If the claimed ammonia removal percentage is higher, additional measurements are necessary to prove this. A sufficiently high water discharge rate must prevent precipitation of salts, dependent on the acid that is used for protonation of the ammonia. In The Netherlands only sulphuric acid may be used for this application. Proper abduction of the water discharge however, remains a problem.

References

Aarnink, A.J.A. and A. Elzing, 1998. Dynamic model for ammonia volatilization in housing with partially slatted floors, for fattening pigs. Livestock production science 53: 153-169.

Aarnink, A.J.A., W.J.M. Landman, R.W. Melse and T.T.T. Huynh, 2005. Systems for eliminating pathogens from exhaust air of animal houses. Livestock environment VII, Beijing, China.

Aarnink, A.J.A., J.W. Schrama, M.J.W. Heetkamp, J. Stefanowska and T.T.T. Huynh, 2006. Temperature and body weight affect fouling of pig pens. Journal of animal science 84: 2224-2231.

Aarnink, A.J.A., A.J. Van den Berg, A. Keen, P. Hoeksma and M.W.A. Verstegen, 1996. Effect of slatted floor area on ammonia emission and on the excretory and lying behaviour of growing pigs. Journal of agricultural engineering research 64: 299-310.

Aarnink, A.J.A. and M.W.A. Verstegen, 2007. Nutrition, key factor to reduce environmental load from pig production. Livestock sciences (accepted).

Bakker, G.C.M. and M.C.J. Smits, 2002. Dietary factors are additive in reducing in vitro ammonia emission from pig manure. Journal of animal science 79 (Suppl. 1): Abstract 757.

Bannink, A., H. Valk and A.M. Van Vuuren, 1999. Intake and excretion of sodium, potassium and nitrogen and the effects on urine production by lactating dairy cows. Journal of dairy science 82: 1008-1018.

Bleijenberg, R., W. Kroodsma and N.W.M. Ogink, 1995. Techniques for the reduction of ammonia emission from a cubicle house with slatted floor (In Dutch). Report 94-35. IMAG, Wageningen, The Netherlands.

Braam, C.R., J.J.M.H. Ketelaars and M.C.J. Smits, 1997a. Effects of floor design and floor cleaning on ammonia emission from cubicle houses for dairy cows. Netherlands journal of agricultural science 45: 49-64.

Braam, C.R., M.C.J. Smits, H. Gunnink and D. Swierstra, 1997b. Ammonia emission from a double - sloped solid floor in a cubicle house for dairy cows. Journal of agricultural engineering research 68: 375-386.

Bussink, D.W., 1996. Ammonia volatilization from intensively managed dairy pastures. Thesis Wageningen University, Wageningen, The Netherlands, 177 pp.

Canh, T.T., A.J.A. Aarnink, Z. Mroz, A.W. Jongbloed, J.W. Schrama and M.W.A. Verstegen, 1998a. Influence of electrolyte balance and acidifying calcium salts in the diet of growing-finishing pigs on urinary pH, slurry pH and ammonia volatilisation from slurry. Livestock production science 56: 1-13.

Canh, T.T., A.J.A. Aarnink, J.B. Schutte, A.L. Sutton, D.J. Langhout, M.W.A. Verstegen and J.W. Schrama, 1998b. Dietary protein affects nitrogen excretion and ammonia emission from slurry of growing-finishing pigs. Livestock production science 56, 181-191.

Canh, T.T., J.W. Schrama, A.J.A. Aarnink, M.W.A. Verstegen, C.E. Van 't Klooster and M.J.W. Heetkamp, 1998c. Effect of dietary fermentable fibre from pressed sugar-beet pulp silage on ammonia emission from slurry of growing-finishing pigs. Animal science 67: 583-590.

Canh, T.T., A.L. Sutton, A.J.A. Aarnink, M.W.A. Verstegen, J.W. Schrama and G.C.M. Bakker, 1998d. Dietary carbohydrates alter the faecal composition and pH and ammonia emission from slurry of growing pigs. Journal of animal science 76: 1887-1895.

Canh, T.T., M.W.A. Verstegen, N.B. Mui, A.J.A. Aarnink, J.W. Schrama, C.E. Van 't Klooster and N.K. Duong, 1999. Effect of nonstarch polysaccharide-rich by-product diets on nitrogen excretion and nitrogen losses from slurry of growing-finishing pigs. Asian-Australian journal of animal science 12: 573-578.

De Bode, M.J.C., 1991. Odour and ammonia emissions from manure storages. In: Odour and ammonia emission from livestock farming. V.C. Nielsen, J.H. Voorburg and P. L' Hermite, Eds. Elsevier Applied Science, London and New York. pp. 59-66.

De Boer, I.J.M., M.C.J. Smits, H. Mollenhorst, G. Van Duinkerken and G.J. Monteny, 2002. Prediction of ammonia emission from dairy barns using feed characteristics. Part I: Relation between feed characteristics and urinary urea concentration. Journal of dairy science 85: 3382-3388.

Ellen, H.H., J. Van Harn and T. Veldkamp, 2005. Desk study on possibilities to reduce ammonia emission from broiler houses (In Dutch). Praktijk Rapport Pluimvee Report 16. Animal Sciences Group, Lelystad, The Netherlands.

Elzing, A. and G.J. Monteny, 1997. Modelling and experimental determination of ammonia emissions rates from a scale model dairy-cow house. Transactions of the ASAE 40: 721-726.

Groenestein, C.M., 2006. Environmental aspects of improving sow welfare with group housing and straw. Thesis Wageningen University, Wageningen, The Netherlands, 146 pp.

Groot Koerkamp, P.W.G., J.H.M. Metz, G.H. Uenk and C.M. Wathes, 1998. Concentrations and emissions of ammonia in livestock buildings in northern Europe. Journal of agricultural engineering research 70: 79-95.

Hacker, R.R., J.R. Ogilvie, W.D. Morrison and F. Kains, 1994. Factors affecting excretory behavior of pigs. Journal of animal science 72: 1455-1460.

Häußermann, A., E. Hartung and T. Jungbluth, 2005. Environmental effects of pig house ventilation controlled by animal activity and CO_2 indoor concentration. In: Precision livestock farming '05. S. Cox, ed. Wageningen Academic Publishers, Wageningen, pp. 57-64.

Hol, J.M.G. and I.H.G. Satter, 1998. Praktijkonderzoek naar de ammoniakemissie van stallen XIX. Vleesvarkensstal met gereduceerd emitterend oppervlak door aangepaste hokinrichting. DLO Report 98-1001. IMAG, Wageningen, The Netherlands.

Huff, W.E., 1984. Effect of litter treatment on broiler performance and certain litter quality parameters. Poultry science 63: 2167-2171.

Huijsmans, J.F.M., 1999. Manure application. In: Monitoring of national ammonia emissions from agriculture. Towards an improved calculation methodology. J.H.A.M. Steenvoorden, Ed. DLO-Staring centrum. Reeks Milieuplanbureau 6, Wageningen, The Netherlands. pp. 139.

Huijsmans, J.F.M., 2003. Manure application and ammonia volatilization. Thesis Wageningen University, Wageningen, The Netherlands, 160 pp.

Huijsmans, J.F.M., J.G.L. Hendriks and G.D. Vermeulen, 1998. Draught requirement of trailing-foot and shallow injection equipment for applying slurry to grassland. Journal of agricultural engineering research 71: 347-356.

Huijsmans, J.F.M. and R.M. De Mol, 1999. A model for ammonia volatilization after surface application and subsequent incorporation of manure on arable land. Journal of agricultural engineering research 74: 73-82.

Huijsmans, J.F.M., J.M.G. Hol and M.M.W.B. Hendriks, 2001. Effect of application technique, manure characteristics, weather and field conditions on ammonia volatilisation from manure applied to grassland. Netherlands journal of agricultural science 49: 323-342.

Huijsmans, J.F.M., J.M.G. Hol and G.D. Vermeulen, 2003. Effect of application method, manure characteristics, weather and field conditions on ammonia volatilization from manure applied to arable land. Atmospheric environment 37: 3669-3680.

Huynh, T.T.T., A.J.A. Aarnink, H.A.M. Spoolder, B. Kemp and M.W.A. Verstegen, 2004. Effects of floor cooling during high ambient temperatures on the lying behavior and productivity of growing finishing pigs. Transactions of the ASAE 47: 1773-1782.

Kroodsma, W., J. Arkenbout and J.A. Stoffers, 1985. New system for drying poultry manure in belt batteries. Report IMAG, Wageningen, The Netherlands.

Melse, R.W. and H.C. Willers, 2004. Toepassing van luchtbehandelingstechnieken binnen de veehouderij. Fase 1: Techniek en kosten. Report Agrotechnology and Food Innovations, Wageningen, The Netherlands.

Monteny, G.J., 2000. Modelling of ammonia emissions from dairy cow houses. Thesis Wageningen University, Wageningen, The Netherlands, 155 pp.

Monteny, G.J. and J.W. Erisman, 1998. Ammonia emission from dairy cow buildings: a review of measurement techniques, influencing factors and possibilities for reduction. Netherlands journal of agricultural science 46: 225-247.

Monteny, G.J., J. Huis in 't Veld, G. Van Duinkerken, G. André and F. Van der Schans, 2001. Towards ammonia emission factors per year for dairy cow housing systems (in Dutch). IMAG Report 2001-09. IMAG, PV and CLM, Wageningen, The Netherlands.

Monteny, G.J., D.D. Schulte, A. Elzing and E.J.J. Lamaker, 1998. A conceptual model for the ammonia emission from free stall cubicle dairy cow houses. Transactions of the ASAE 41: 193-201.

Monteny, G.J., M.C.J. Smits, G. Van Duinkerken, H. Mollenhorst and I.J.M. De Boer, 2002. Prediction of ammonia emission from dairy barns using feed characteristics. Part II: Relation between urinary urea concentration and ammonia emission. Journal of dairy science 85: 3389-3394.

Ogink, N.W.M. and W. Kroodsma, 1996. Reduction of ammonia emission from a cow cubicle house by flushing with water or a formalin solution. Journal of agricultural engineering research 63: 197-204.

Reitsma, B. and C.M. Groenestein, 1995. Praktijkonderzoek naar de ammoniakemissie van stallen XIX. Hellingstal voor vleesvarkens. DLO Report 95-1002. DLO, Wageningen, The Netherlands.

Scholtens, R., 1996. Inspectie van luchtwassystemen voor mechanisch geventileerde varkensstallen. Report IMAG, Wageningen, The Netherlands.

Sinnot, R.K., 1999. Coulson & Richardson's chemical engineering. Butterworth-Heinemann, Oxford, England.

Smits, M.C.J., 1998. Grooving a solid V-shaped floor; some observations on walking behaviour and ammonia emission (in Dutch). IMAG Report P 98-60. IMAG-DLO, Wageningen, The Netherlands.

Smits, M.C.J., G.J. Monteny and G. Van Duinkerken, 2003. Effect of nutrition and management factors on ammonia emission from dairy cow herds: models and field observations. Livestock production science 84: 113-123.

Smits, M.C.J., H. Valk, A. Elzing and A. Keen, 1995. Effect of protein nutrition on ammonia emission from a cubicle house for dairy cattle. Livestock production science 44, 147-156.

Smits, M.C.J., H. Valk, G.J. Monteny and A.M. Van Vuuren, 1997. Effect of protein nutrition on ammonia emission from cow houses. In: Gaseous nitrogen emissions from grasslands. S.C. Jarvis and B. Pain, Eds. CAB International, Wallingford, England.

Sutton, A.L., K.B. Kephart, M.W.A. Verstegen, T.T. Canh and P.J. Hobs, 1999. Potential for reduction of odorous compounds in swine manure through diet modification. Journal of animal science 77: 430-439.

Swierstra, D., C.R. Braam and M.C.J. Smits, 2001. Grooved floor system for cattle housing: Ammonia emission reduction and good slip resistance. Applied engineering in agriculture 17: 85-90.

Swierstra, D., M.C.J. Smits and W. Kroodsma, 1995. Ammonia emission from cubicle houses for cattle with slatted and solid floors. Journal of agricultural engineering research 62: 127-132.

Van der Hoek, K.W., 2002. Input variables for manure and ammonia data in the Environmental Balance 2001 and 2002 including dataset agricultural emissions 1980-2001 (in Dutch). RIVM Report 773004013. RIVM, Bilthoven, The Netherlands.

Van der Molen, J., A.C.M. Beljaars, W.J. Chardon, W.A. Jury and H.G. Van Faassen, 1990a. Ammonia volatilization from arable land after application of cattle slurry. 2. Derivation of a transfer model. Netherlands journal of agricultural science 38: 239-254.

Van der Molen, J., H.G. Van Faassen, M.Y. Leclerc, R. Vriesema and W.J. Chardon, 1990b. Ammonia volatilization from arable land after application of cattle slurry. 1. Field estimates. Netherlands journal of agricultural science 38: 145-158.

Van Duinkerken, G., G. André, M.C.J. Smits, G.J. Monteny, K. Blanken, M.J.M. Wagemans and L.B.J. Sebek, 2003. Relationship between diet and ammonia emissions from a dairy cow house (in Dutch). PV Report 25. PV / IMAG, Wageningen, The Netherlands.

Van Duinkerken, G., G. André, M.C.J. Smits, G.J. Monteny and L.B.J. Sebek, 2005. Effect of rumen-degradable protein balance and forage type on bulk milk urea concentration and emission of ammonia from dairy cow houses. Journal of dairy science 88: 1099-1112.

Van Vuuren, A.M. and M.C.J. Smits, 1997. Effect of nitrogen and sodium chloride intake on production and composition of urine in dairy cows. In: Gaseous nitrogen emissions from grasslands. S.C. Jarvis and B. Pain, Eds. CAB International, Wallingford, England.

VROM, 2006. Regeling ammoniak en veehouderij. Staatscourant 207 (October).

Williams, A.G. and E. Nigro, 1997. Covering slurry stores and effects on emission of ammonia and methane. In: International symposium on ammonia and odour control from animal production facilities. J.A.M. Voermans and G.J. Monteny, Eds., PV Rosmalen. pp. 421-428.

6. Measurement methods and strategies

Julio Mosquera Losada

In order to monitor the efficiency of emission reducing techniques applied in agriculture, techniques should be available to accurately measure the ammonia emission from all different agricultural sources. Each ammonia source has specific properties, depending on factors that influence the emission rate from that particular source (see Chapter 3). The choice of measuring method should therefore be suitable for a particular source.

This chapter describes the measurement methods and strategies used in The Netherlands to measure the ammonia emissions from animal houses. Some alternatives are also suggested. The selection of measurement methods is dependent on different factors, including the purpose of the measurements, the properties of the selected source category, and the possibility of measuring other compounds (greenhouse gases, odour, dust) when measuring ammonia emissions. These factors are described in Section 6.1. Section 6.2 shortly describes the current measurement protocol (method + strategy) used in The Netherlands to measure ammonia emissions from agriculture. Based on results of the analysis of data series of ammonia emission measurements, new measurement protocols will be suggested (Section 6.3 and 6.4).

6.2. What and how do we want to measure?

6.1.1. Goal of the measurements

The selection of a specific measurement protocol to measure agricultural ammonia emissions depends on different factors. First of all, the purpose of the measurements should be clearly defined. In this way, the requisites that have to be fulfilled during the measurements can be specified. In addition, it is important to know beforehand whether additional compounds should be measured. Although there are different users, which also have different goals for the measurements, the following objectives can be defined:
1. Assessment/control of an emission factor. These factors are determined by measuring the ammonia emissions from a specific source according to standardised measurement protocols, which specify the measurement technique, the duration and the conditions (including specific zootechnical values) of the measurements.

2. Monitoring ammonia emissions for legislation purposes, to verify the assigned emission factors.
3. Monitoring ammonia emissions to investigate and better understand the processes behind ammonia emissions. This requires measurements with a high time resolution in order to follow the dynamics of the ammonia emission, and to relate the emission pattern with underlying processes affecting the emission.
4. Management support, by coupling measurement techniques for ammonia concentration with the computer system used in the farm. The objective is to provide the farmer with a tool to visualise how changes in farm management influence the ammonia concentrations in the animal house.

These different objectives will result in a different approach and related different prerequisites for the measurements. Parameters of importance include the measurement period, frequency of data collection, accuracy, representativeness and costs of the measurements (Hofschreuder *et al.*, 2003).

6.1.2. Agricultural emission sources

Once the purpose of the measurements is clear, the second point of consideration is to define the ammonia emission source where measurements should be performed. The following source categories can be defined:
1. Mechanically ventilated animal houses.
2. Naturally ventilated animal houses, with small inlet openings.
3. Naturally ventilated animal houses, with large inlet openings.
4. Animal houses (naturally or forced ventilated) with outdoor yards.
5. Manure storages outside the animal houses.
6. The application of manure onto arable land or grassland (see Section 5.4).
7. Grazing (see Section 5.1).

Because each emission source has unique characteristics, the selection of the measurement equipment and the representativeness of the measurements will also differ for different sources. For example, animal houses are, in general, a well-defined source, with the emission occurring from a confined space. But while in mechanically ventilated animal houses (Figure 6.1A) the inlets and outlets are, in general, well defined, this is usually not the case for naturally ventilated animal houses (Figure 6.1B), because openings can act

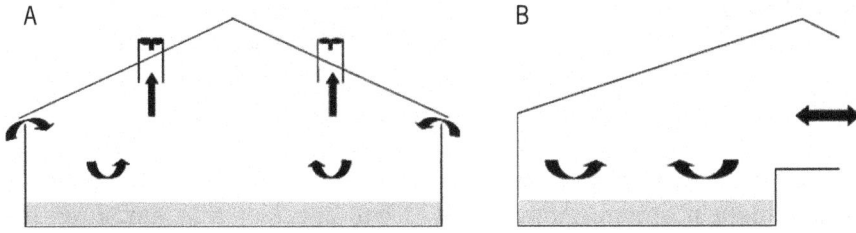

Figure 6.1. Schematic representation of the flow pattern for a particular mechanically ventilated (A) and naturally ventilated (B) animal house.

in some cases as inlet and in some other cases as outlet, depending on the weather conditions (e.g. wind speed and wind direction).

6.1.3. Compounds to be measured

Depending on the purpose of the measurements, it can be decided to measure only one specific compound (in this case ammonia) or to choose for an integral approach and measure simultaneously a combination of different compounds. The costs involved in the measurements are the main disadvantage of using an integral approach. As advantage, the possibility to investigate the effect of reducing the emission of one compound, on the emissions of the other compounds. In general, a selection (or combination) will be made between the following compounds:
- ammonia (NH_3);
- greenhouse gases (methane (CH_4); nitrous oxide (N_2O); carbon dioxide (CO_2));
- odour;
- dust.

6.2. Current measurement protocols

6.2.1. Animal houses

The Dutch government uses emission factors as a tool to determine ammonia emissions from animal houses and to assess the impact of reduction policies. These emission factors are based on measurements for a particular animal category and housing system according to standardised measurement protocols (Anonymous, 1993; Anonymous, 1996), which specify the measurement technique, the duration and the conditions of the measurements (e.g. periods of the year where measurement should be performed, management, zootechnical values, and more).

To measure the ammonia emission from an animal house, the current measurement protocol (Anonymous, 1996) specifies that measurements should be performed of the air exchange rate (or ventilation rate) and of the ammonia concentration in the air leaving the animal house. In mechanically ventilated animal houses the air leaves the house mainly through the ventilation shafts. The ammonia concentration is therefore sampled in the ventilation shafts (Figure 6.2A), and transported through an isolated and heated sampling line to the measurement equipment. The current measurement protocol specifies two techniques to measure the ammonia concentration: (1) an NO_x monitor used after a necessary conversion of NH_3 to NO in a dedicated catalytic converter, and (2) a photo-acoustic infrared monitor. For the ventilation rate, commercially available fan wheel anemometers are specified for measurements in mechanically ventilated animal houses. They should be placed under the ventilator and cover the whole ventilation shaft area (Figure 6.2B). The ammonia emission is then calculated as the product of the measured ammonia concentration and ventilation rate (Demmers *et al.*, 1999; Groenestein, 1993; Groenestein *et al.*, 2001; Groot Koerkamp *et al.*, 1998; Phillips *et al.*, 1998; Scholtens, 1993).

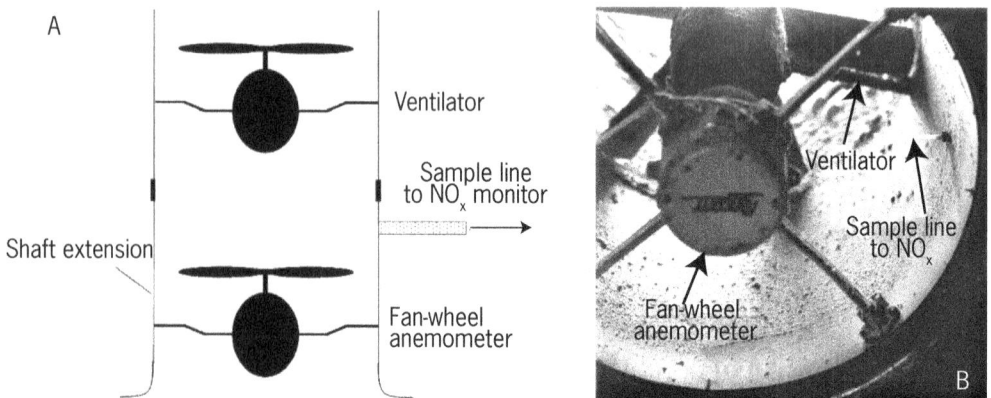

Figure 6.2. Schematic (A) and 'in-situ' application (B) of the fan wheel anemometer and the sampling line to the NO_x monitor for ammonia emission measurements in mechanically ventilated animal houses.

For naturally ventilated animal houses, the internal tracer gas ratio method (mass balance approach) is widely used (Huis in 't Veld and Groot Koerkamp, 2001; Huis in 't Veld and Monteny, 2003; Huis in 't Veld *et al.*, 2001; Monteny *et al.*, 2005; Monteny *et al.*, 2002; Mosquera and Ogink, 2004; Snell *et al.*, 2003; Stout and Richard, 2003; Zhang *et al.*, 2005). In this method, a tracer gas is introduced in the animal house at a constant rate (Q_{tracer}). It is important that the tracer gas is injected in such a way that it mixes well with other gases (e.g. ammonia) in the animal house. Therefore, the tracer gas should be injected close to the ammonia sources (Figure 6.3A). In addition, the tracer gas should be easy detectable with the used equipment, inert, and with low background concentrations. For these reasons, sulphur hexafluoride (SF_6) is commonly used as tracer gas. However, other tracer gases could be used as well.

The second step within the internal tracer gas ratio method is to measure the concentrations of ammonia (C_{NH3}) and the tracer gas (C_{tracer}) at a place where a representative sample could be obtained. That means that the residence time for both gases should be large enough to allow the gases to mix properly. For very open animal houses (or naturally ventilated animal houses with large openings), residence time could be a problem when using the internal tracer gas ratio method. In general, a sampling line with a number of sampling points is placed under the ridge in the animal house (Figure 6.3B), to obtain an average concentration of both gases along the animal house.

Figure 6.3. Injection system (A) and sampling line for ammonia and tracer gas concentrations (B) for ammonia emission measurements in naturally ventilated animal houses.

The ammonia emission (Q_{NH3}) is then calculated using the following equation:

$$Q_{NH_3} = \frac{Q_{tracer}}{C_{tracer}} \times C_{NH_3} \qquad (1)$$

Although the current measurement protocol does not prescribe any specific measurement method, the measurement equipment (NO_x monitor + converters, or photo acoustic infrared monitor) used for measurements at mechanically ventilated animal houses is usually applied to measure the ammonia concentration in naturally ventilated animal houses. To measure the tracer gas (SF_6) concentrations, a gas chromatograph equipped with an Electron Capture Detector (GC-ECD) is commonly used.

6.2.2. Other sources

There are no specific measurement protocols in The Netherlands to measure the ammonia emission from outside manure storage facilities.

6.3. New measurement protocols

An important characteristic of the current measurement protocol for animal houses is that ammonia concentrations and ventilation rates are measured continuously, or at least 1-hour average values should be provided. This provides a large amount of information about the emission process, and the daily and seasonal patterns of the measured emission. However, costs of these measurements on equipment, maintenance and labour are high. And due to these high costs, the number of animal houses that can be measured is usually quite small. In order to reduce the costs of the measurements, without sacrificing the quality of the emission estimate, two different approaches can be used:

• New sampling strategies, based on data series and statistical analysis: determination of the different sources of variance associated with the measurements.
• New measurement techniques: accuracy, costs, and how representative is the result obtained with these new techniques in relation with the current measurement methods.

To define new strategies, the different sources of variance of the measurements should be clearly identified:

- the variation within a farm;
- the variation between farms;
- the variation associated with the used measurement equipment.

The current measurement protocol for the determination of ammonia emission factors specifies that measurements should be performed in one animal house, which is considered to be representative for the housing system being measured. In this way, the variation between farms is considered to be negligible. However, previous studies (Mosquera and Ogink, 2004, 2006) have shown that management is also an important factor, and that differences between animal houses with the same housing system could be larger than the variation in time and season within one animal house. Therefore, more locations should be measured in order to minimise the variation between farms. Similar conclusions have been found for odour emissions (Ogink and Klarenbeek, 1997). Besides, due to autocorrelation, not all measurements are independent, which underlines the possibility of reducing the number of measurements by using shorter measurement periods, but distributed over the whole year (Vranken *et al.*, 2004).

The variation associated with the measurement equipment is usually considered negligible with respect to other sources of variance, if enough measurements (within or between different locations) or replicates are performed (Mosquera and Ogink, 2006).

To summarise, in order to increase the accuracy of measurements without simultaneously increasing the costs, the following recommendations are made:
1. Increase the number of measurement locations.
2. Reduce the number of measurements per location.
3. Use, if possible, cheaper measurement equipment, as long as systematic errors are avoided.
4. Because some factors affecting the ammonia emission from animal houses follow a diurnal pattern, use of daily averages is highly recommended.

6.4. Alternative measurement methods

An extended review of the different existing methods for measuring the concentrations, ventilation rate (in animal houses) and emissions can be found in literature (Arogo *et al.*, 2001; Hofschreuder *et al.*, 2003; Mosquera *et al.*, 2002a, 2005; Ni and Heber, 2001; Phillips *et al.*, 2000; Scholtens, 1993; Van 't Klooster *et al.*, 1992). This section summarises the most promising alternative methods for a number of specific ammonia sources.

6.4.1. Mechanically ventilated animal houses

Passive flux samplers
Passive flux samplers (PFS) represent an alternative that allows the monitoring of ammonia emissions from a larger number of animal houses at a low price (Monteny *et al.*, 1999; Mosquera, 2003, 2002b; Scholtens *et al.*, 2003a,b). Passive flux samplers (Figure 6.4) collect ammonia at a rate proportional to the wind velocity of the air stream passing it without the need of a pump or other instrument requiring power supply. In this way, wind speed or ventilation rate measurements are not necessary. In order to determine this proportional (constant) factor, the sampler should be first calibrated in a wind tunnel, by measuring the pressure difference through the sampler for different wind speeds. Passive flux samplers are easy to construct, transport and handle, and have little lab requisites. However, similar to other passive sampling methods, passive flux samplers are based on the principle of diffusion to a reaction surface and, therefore, need longer sampling periods. For animal houses, where the ammonia concentrations are usually high, an average time of 4 hours is enough, although larger averaging periods (~1 week) are commonly used. It is important to point out that not all mechanically ventilated animal houses are suitable for the use of passive flux samplers. The flow around the sampler should be oriented (as much as possible) in the direction of the sampler. In addition, existing obstacles should not change the pressure difference along the sampler. When use of PFS is possible, the samplers are placed in or under the ventilation shafts. In order to assure a good alignment of the sampler with respect to the flow, there should be a sufficient separation distance between the samplers and the ventilator. A duplicate is usually advised for all measurements (Figure 6.4).

Ammonia absorbing devices
Another alternative to the current measurement protocol is using simple (and cheap) ammonia absorbing devices (Willems badges, impingers, denuders) to measure the ammonia concentration in the animal house. However, these measurements have to be complemented with ventilation rate measurements. An alternative for the fan wheel anemometers (mechanically ventilated animal houses) and the internal tracer gas ratio method (naturally ventilated animal houses with small openings) is the CO_2

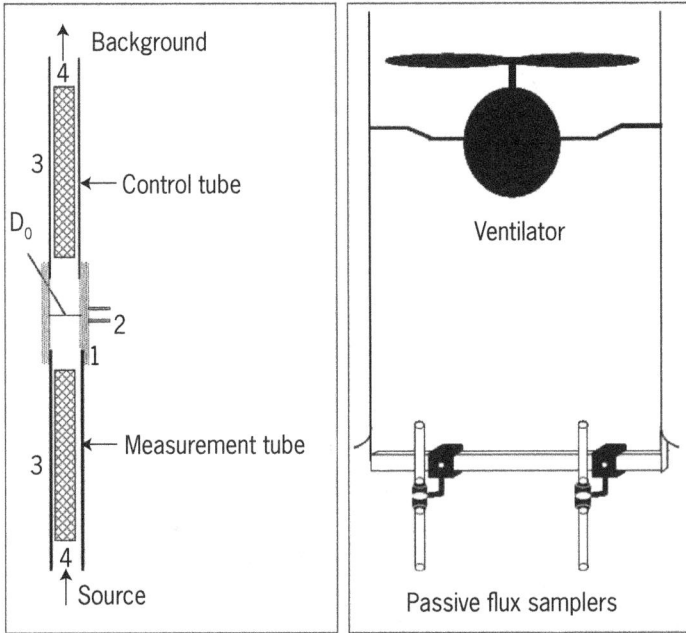

Figure 6.4. Schematic drawing of the ammonia passive flux sampler and application under a ventilation shaft. (1) Connecting piece with orifice; (2) Pressure points (optional); (3) Sampling tube; (4) Absorption medium (for measurements in animal houses, filter paper strip loaded with sulphuric acid); D_o, orifice diameter.

mass balance method. In this method, the CO_2 concentration inside the animal house is measured and compared with the concentration expected by taking into account the CO_2 production from the animals. This method has been applied to determine the ventilation rate from both mechanically ventilated (Blanes and Pedersen, 2005; Li *et al.*, 2004; Pedersen *et al.*, 1998; Xin *et al.*, 2006) as from naturally ventilated (Van 't Klooster and Heitlager, 1994; Zhang *et al.*, 2005) animal houses.

Open-path systems
Another method to measure ammonia concentrations is to use an open-path system (e.g., Tuneable Diode Laser (TDL), Fourier Transformed InfraRed (FTIR), or Differential Optical Absorption Spectroscopy (DOAS)). By placing the instrument (detector) in one side of the animal house, and a reflector in the other side, it is possible to obtain an average ammonia concentration along the animal house. Open-path systems have already

been used to measure ammonia concentrations in the open field (Amon *et al.*, 1997; Coates *et al.*, 2004; Flesch *et al.*, 2004, 2005; Harris *et al.*, 2001; Monteny *et al.*, 2005; Secrest, 2000; Sommer *et al.*, 2004). Further research is necessary to identify possible interferences due to other gases present in the animal house, and the stability of the laser under practical conditions. This method also needs simultaneous ventilation rate measurements in order to obtain an ammonia emission rate.

Open path measurements have also been performed using Laser Imaging Detection And Ranging (LIDAR). Instead of a reflector, this technique makes use of the back scattering of laser light. Changes in the received light arise from partial absorption by individual aerosols and even molecules such as ammonia. Additional Doppler measurements provide the necessary flux data to calculate source emission values (Zhao *et al.*, 2002).

6.4.2. Naturally ventilated animal houses with small openings

One alternative is to use the internal tracer gas ratio method in combination with passive samplers (Willems badges) or impingers. Using these techniques has the advantage of reducing the installation costs when compared with measurements using the NO_x monitor or the photo acoustic infrared monitor. However, they only provide an average emission over the whole measurement period, and are therefore not useful when the purpose of the measurements is to study the dynamics of the ammonia emission process. The internal tracer gas ratio method could also be combined with an open-path system. However, as already mentioned in Section 6.4.1, this method has not been tested yet.

6.4.3. Naturally ventilated animal houses with large openings

In some cases such as naturally ventilated animal houses with large openings, mixing of the tracer gas is compromised. In such cases, ammonia emission measurements should be performed outside the animal house. One possibility is to use a micrometeorological mass balance approach (integrated horizontal flux, IHF). In this method, a set of masts is placed downwind of the source (Figure 6.5). The measurement equipment (e.g., passive flux samplers (Mosquera *et al.*, 2002b; Sommer *et al.*, 2004), or a combination of anemometers for meteorological parameters, and Willems badges or denuders for ammonia concentrations (Mosquera *et al.*, 2002b)) is installed at different heights in every mast. The difference between the

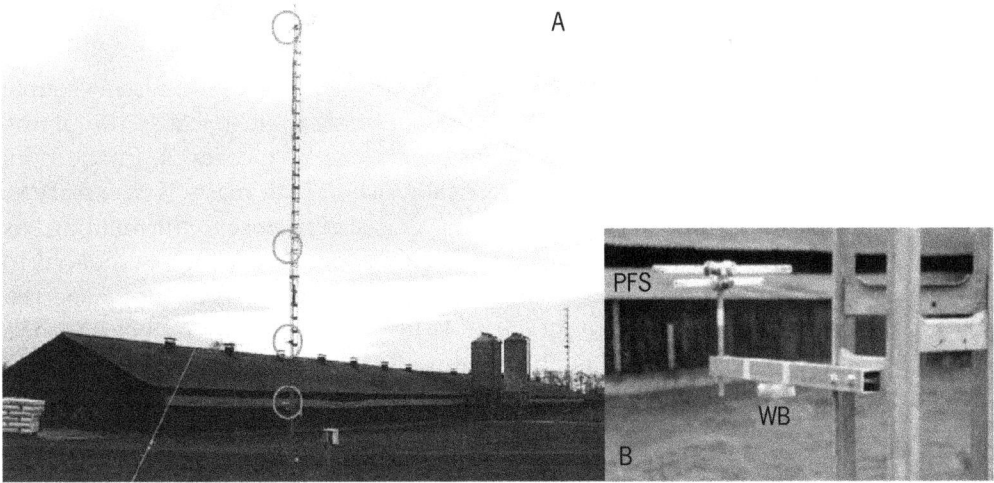

Figure 6.5. (A) Flux frame method applied in a naturally ventilated animal house for poultry. (B) passive sampling techniques (WB = Willems badges; PFS = passive flux samplers).

horizontal fluxes through the vertical planes upwind and downwind of the source gives the emission from the animal house:

$$E_{NH_3} = \frac{1}{x} \times \sum_{i=1}^{i=n} u_i \times (c_i^d - c_i^u) \times z_i \cdot A \qquad (2)$$

E_{NH_3} = ammonia emission from the animal house [$\mu g.s^{-1}$]
i = height of the mast [m] where measurements are performed
n = number of heights in the mast
j = mast where measurements are performed
m = number of masts
u_i = wind speed [$m.s^{-1}$] at height i
c_i^d = ammonia concentration [$\mu g.m^{-3}$] at height i downwind of the
       source
c_i^u = ammonia concentration [$\mu g.m^{-3}$] at height i upwind of the source
z_i = height [m] where measurements are considered to be
       representative
x = fetch [m]
A = surface area [m^2] of the emitting source

This method has the advantage of not interfering with the gas emission and exchange processes between the emitting surface and the atmosphere. However, it requires simultaneous measurements of wind speed and ammonia concentrations at the same height in order to calculate the horizontal ammonia flux for that particular measurement point. Moreover, the whole ammonia plume has to be measured. Far from the emission source (animal house), the ammonia plume is uniform, but because the plume is rapidly dispersed, lower concentrations will be measured. In addition, since the plume also disperses vertically, use of high masts is necessary to capture the complete ammonia plume. Measuring close to the building has as advantage of having concentrations that are usually high compared to background concentrations. Moreover, because the plume is not dispersed vertically, short masts are sufficient. However, the flow pattern close to the animal house is affected by the building.

Another option is to use the external tracer gas ratio method (Kaharabata *et al.*, 2000; Mosquera *et al.*, 2002a), which uses the same principle described in Section 6.2.1 for naturally ventilated animal houses with small openings (internal tracer gas ratio method). In this method, a tracer gas is released from the source (in this case the animal house) at a know rate (E_{tracer}) in such a way that it mixes well with the emitted gases (such as ammonia or greenhouse gases). Then the concentrations of the tracer gas (C_{tracer}) and of the gas to be measured (e.g. ammonia; C_{NH_3}) are measured downwind of the emitting source. Assuming that both the tracer gas and ammonia disperse in the same way, the ammonia emission from the source can be calculated from the ratio of the measured concentrations and the injection rate of the tracer gas according to the following equation:

$$E_{NH_3} = E_{tracer} \times \frac{C_{NH_3}}{C_{tracer}} \tag{3}$$

A different approach is to use dispersion models and downwind ammonia concentration measurements to determine the ammonia emission from the animal house. One of such models is the Gaussian plume dispersion model (Czepiel *et al.*, 1996; Hensen and Scharff, 2001; Mosquera *et al.*, 2002a; Scharff and Hensen, 1999). This model assumes a Gaussian concentration distribution in horizontal and vertical directions downwind from the source. This model calculates the contribution of a single source to a certain receptor point according to the equation:

$$C(x,y,z) = \frac{Q}{2\pi \times u \times \sigma_y \times \sigma_z} \times e^{-y^2/(2\sigma_y^2)} \times \left(e^{-(z-H)^2/(2\sigma_z^2)} + e^{-(z+H)^2/(2\sigma_z^2)} \right) \tag{4}$$

In this equation, $C(x,y,z)$ is the downwind air concentration due to a continuous source of constant strength Q located at the point (0,0,H), where H is the height of the emission source. The coordinates x, y, z are oriented, respectively, in the direction of the mean wind u, horizontal and normal to u, and vertical and normal to u. The dispersion parameters σ_y and σ_z, which are dependent on the stability of the atmosphere during the measurements, can be calculated empirically by using micrometeorological measurements or a tracer gas (Chen *et al.*, 1997). The Gaussian plume model is not designed for situations where buildings may have a dominant influence on the dispersion process, which is the case close to buildings (<100 m). Besides, the dispersion parameters σ_y and σ_z are not valid for distances less than 100 m.

A second modelling approach is represented by the backwards Lagrangian Stochastical (bLS) model (Flesch *et al.*, 1995, 2004, 2005; Sommer *et al.*, 2004). This model calculates trajectories of air parcels (or particles) upwind from the receptor point and backwards to the emitting source. It requires only one-point or line measurements of wind speed and concentration (C) within the downwind plume to calculate the emission strength (Q) of the emitting surface:

$$Q = \frac{(C-C_b)}{(C/Q)_{sim}}; (C/Q)_{sim} = \frac{1}{N} \sum \left| \frac{2}{w_0} \right| \tag{5}$$

In this equation, N is the total number of air parcels (or particles) released from the receptor point, w_0 is the vertical velocity of the air parcel when arriving (touching) the emitting source, and C_b is the background concentration. Only those air parcels arriving to the source are taken into account in the calculation.

6.4.4. Animal houses with outdoor yards

For animal houses with outdoor yards, two different approaches are possible:
1. To measure the complete system (animal house + outdoor) as a whole.
2. To measure the animal house and the outdoor yards independently.

When selecting the first approach, measurements should be performed outside the whole farming system. The methods described in Section 6.4.3 (integrated horizontal flux, tracer gas ratio method, Gaussian plume model, backwards Lagrangian Stochastical) are also suitable for these measurements. This approach should also be applied when dealing with outdoor yards by naturally ventilated animal houses with large openings. Since the emission from the animal house has to be measured outside the house, it is not possible to differentiate the contribution from the animal house and the outside yard independently, and the system has to be considered as a whole.

The selection of the second approach requires a method to measure the emissions from the animal house, and another method to measure the emission from the outdoor yards. The emissions from the animal house can be measured by using one of the methods described in Sections 6.2.1 (current measurement protocol), 6.4.1 (alternative methods for mechanically ventilated animal houses) and 6.4.2 (alternative methods for naturally ventilated animal houses with small openings). To measure the emission from outdoor yards only, enclosure techniques are usually applied. Two different types of enclosure techniques can be defined: static (closed) and dynamic (open) chambers. They are usually inserted into the ground or placed on top of frames, previously installed at the measurement location. In static chambers, air inside the chamber is usually re-circulated, to allow for a good mixing of all gases in the headspace of the chamber. Static chambers (Figure 6.6A) use the law of conservation of mass to determine the gas exchange between the soil surface and the atmosphere:

$$V \times \frac{dC_i}{dt} = - \phi \times (C_i - C_e) + E_{NH_3} - f \tag{6}$$

In this equation, E_{NH_3} is the emission rate from the surface area covered by the chamber, C_i and C_e are, respectively, the concentrations of the gas measured inside and outside (background) the chamber, ϕ and f represent, respectively, the leakage and adsorption/absorption losses, and V is the volume (headspace) of the chamber.

When the emission and the leakage and adsorption/absorption losses remain constant during the whole measurement period, the concentration of the gas inside the chamber can be determined according to:

$$C_i = \left(\frac{\phi \cdot C_e + E_{NH_3}}{\phi + E_{NH_3}} \right) \times \left(1 - e^{- \frac{\phi + E_{NH_3}}{V} \times t} \right) + C_o \times e^{- \frac{\phi + E_{NH_3}}{V} \times t} \tag{7}$$

In this equation, C_0 represents the gas concentration inside the chamber at time t=0. When leakage and adsorption/absorption losses are considered to be negligible, the increase in gas concentration (C_{t2}-C_{t1}) in the headspace

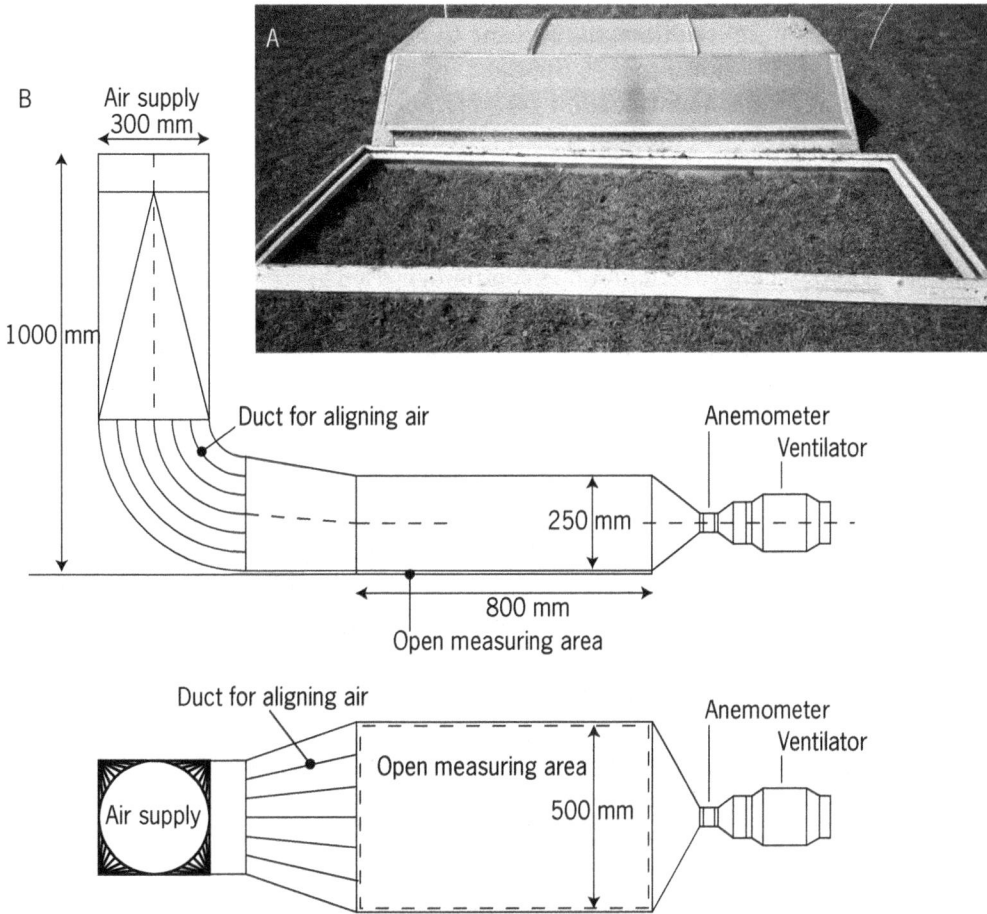

Figure 6.6. Static (A) and dynamic (B) chambers for emission measurements in outdoor yards.

volume (V_c) during the measurement period ($\Delta t = t_2 - t_1$) can be used to determine the emission (E_{NH3}) from the covered emitting surface (A_s):

$$E_{NH_3} = \frac{V_c}{As \times \Delta t} \times (C_{t2} - C_{t1})$$

(8)

Static chambers have the advantage of being cheap and easy to use. However, they can interfere and influence the emission and exchange processes between the emitting surface and the atmosphere. By injecting a tracer gas into the chamber, and applying the internal tracer gas approach to calculate the emission (Hensen *et al.*, 2004), measurements can be performed over a shorter period of time, minimising the (possible) influence of the chamber on the emission process. Spatial variability is also a major limitation for the accurate quantification of emission using static chambers, unless a large number of measurements is performed in different places across the outdoor yard.

In dynamic chambers (or wind tunnels; Figure 6.6B (Aarnink *et al.*, 2002; Ivanova-Peneva *et al.*, 2006); a ventilator is used to pump air at a constant flow rate (Q) from one side of the chamber (inlet) through the emission source to the other side of the chamber (outlet). The difference in concentration between the outlet (C_{out}) and the inlet (C_{in}) is then used to determine the emission strength (E_{NH3}) from the covered area (A_s):

$$E_{NH_3} = \frac{Q}{As} \times (C_{out} - C_{in})$$
(9)

The ammonia concentration is usually measured (both for static and dynamic chambers) either with gas detection tubes, absorption flasks or on-line measurement equipment (FTIR, TDL).

6.4.5. Outside manure storages

Different methods have been applied both in The Netherlands and elsewhere to determine the emissions of ammonia from outside manure storage facilities (Amon *et al.*, 2001; Chadwick, 2005; Hellebrand and Kalk, 2001; Hess and Hügle, 1994; Karlsson, 1996; Monteny *et al.*, 2005; Mosquera *et al.*, 2004; Phillips *et al.*, 1997; Sneath *et al.*, 2006; Sommer *et al.*, 2004). These include:
1. Enclosure techniques (see Section 6.4.4).
2. Integrated Horizontal Flux approach (see Section 6.4.3).
3. External tracer gas ratio method (see Section 6.4.3).
4. Dispersion modeling (see Section 6.4.3).

When manure storages are located in the vicinity of other sources (such as the animal house itself), the application of the external tracer gas ratio method or dispersion modelling is not adequate: with these methods it is not possible to differentiate between the contribution of the storage facility and other sources.

References

Aarnink, A.J.A., M. Wagemans and A. Beurskens, 2002. Development of a procedure to measure local ammonia emissions in organic pig farming (in Dutch). Report 2028. IMAG, Wageningen, The Netherlands.

Amon, B., T. Amon, J. Boxberger and C. Alt, 2001. Emissions of NH_3, N_2O and CH_4 from dairy cows housed in a farmyard manure tying stall (housing, manure storage, manure spreading). Nutrient cycling in agroecosystems 60: 103-113.

Amon, B., J. Boxberger, T. Amon, A. Gronauer, G. Depta, S. Neser and K. Schäfer, 1997. Methods for measuring emissions from agrarian sources: FTIR measurement techniques with White-Cell, large chamber or open-path. In: International symposium ammonia and odour control from animal production facilities. J.A.M. Voermans and G.J. Monteny, Eds., Vinkeloord, The Netherlands.

Anonymous, 1993. Beoordelingsrichtlijn in het kader van Groen Label stallen. LNV / VROM Report LNV / VROM, The Hague, The Netherlands.

Anonymous, 1996. Beoordelingsrichtlijn in het kader van Groen Label stallen (update). LNV / VROM Report LNV / VROM, The Hague, The Netherlands.

Arogo, J., P.W. Westerman, A.J. Heber, W.P. Robarge and J.J. Classen, 2001. Ammonia in animal production. A review. ASAE Annual International Meeting, Sacramento, California, USA.

Blanes, V. and S. Pedersen, 2005. Ventilation flow in pig houses measured and calculated by carbon dioxide, moisture and heat balance equations. Biosystems engineering 92: 483-493.

Chadwick, D.R., 2005. Emissions of ammonia, nitrous oxide and methane from cattle manure heaps: effect of compaction and covering. Atmospheric environment 39: 787-799.

Chen, Y., S.J. Hoff and D.S. Bundy, 1997. Evaluation of models for the dispersion parameters sy and sz. Fifth international symposium livestock environment, Minnesota, USA.

Coates, T., S.M. McGinn and J. Bauer, 2004. Application of open-path TDL analysers for determination of methane and ammonia emissions from livestock facilities. 26th Conference on agricultural and forest meteorology, Vancouver, Canada.

Czepiel, P.M., B. Mosher, C. Harriss, J.H. Shorter, J.B. McManus, C.E. Kolb, E. Allewine and B.K. Lamb, 1996. Landfille methane emissions measured by enclosure and atmospheric tracer methods. Journal of geographical research 101 (D11): 16711-16719.

Demmers, T.G.M., L.R. Burgess, J.L. Short, V.R. Phillips, J.A. Clark and C.M. Wathes, 1999. Ammonia emissions from two mechanically ventilated UK livestock buildings. Atmospheric environment 33: 217-227.

Flesch, T., J.D. Wilson and E. Yee, 1995. Backward-time Lagrangian stochastic dispersion models, and their application to estimate gaseous emissions. Journal of applied meteorology 34: 1320-1332.

Flesch, T.K., J.D. Wilson, L.A. Harper and B.P. Crenna, 2005. Estimating gas emissions from a farm with an inverse-dispersion technique. Atmospheric environment 39: 4863-4874.

Flesch, T.K., J.D. Wilson, L.A. Harper, B.P. Crenna and R.R. Sharpe, 2004. Deducing ground-to-air emissions from observed trace gas concentrations: a field trial. Journal of applied meteorology 43: 487-502.

Groenestein, C.M., 1993. Animal-waste management and emission of ammonia from livestock housing systems: Field studies. Fourth international symposium 'Livestock environment IV', Coventry, England, ASAE.

Groenestein, C.M., J.M.G. Hol, H.M. Vermeer, L.A. den Hartog and J.H.M. Metz, 2001. Ammonia emission from individual- and group-housing systems for sows. Netherlands journal of agricultural science 49: 313-322.

Groot Koerkamp, P.W.G., J.H.M. Metz, G.H. Uenk and C.M. Wathes, 1998. Concentrations and emissions of ammonia in livestock buildings in northern Europe. Journal of agricultural engineering research 70: 79-95.

Harris, D.B., R.C. Shores and L.G. Jones, 2001. Ammonia emission factors from swine finishing operations. 10th Annual international emission inventory conference, one Atmosphere, one Inventory, many challenges, Washington D.C., USA.

Hellebrand, H.J. and W.D. Kalk, 2001. Emission of methane, nitrous oxide, and ammonia from dung windrows. Nutrient cycling in agroecosystems 60: 83-87.

Hensen, A., T. Groot, W.C.M. Van den Bulk, M.J. Blom, P.A.C. Jongejan and A.T. Vermeulen, 2004. Looking for unknown sources of CH_4 and N_2O on farm sites using a fast response box system. Report C-04-067. ECN, Petten, The Netherlands.

Hensen, A. and H. Scharff, 2001. Methane emission estimates from landfills obtained with dynamic plume measurements. Water air and soil pollution 1: 455-464.

Hess, H.J. and T. Hügle, 1994. Vergleichende messung zur NH_3-emission. Landtechnik 49: 362-363.

Hofschreuder, P., J. Mosquera, J.M.G. Hol and N.W.M. Ogink, 2003. Ontwerp van nieuwe meetprotocollen voor het meten van gasvormige emissies in de landbouw. Report 008. Agrotechnology & Food Innovations, Wageningen, The Netherlands.

Huis in 't Veld, J.W.H. and P.W.G. Groot Koerkamp, 2001. Research into the ammonia emission from livestock production systems no. L; Naturally ventilated cubicle-housing system with a profiled floor for dairy cattle in the winter period (in Dutch). Report IMAG, Wageningen, The Netherlands.

Huis in 't Veld, J.W.H. and G.J. Monteny, 2003. Methane emissions from naturally ventilated animal houses for dairy cattle (in Dutch). Report IMAG, Wageningen, The Netherlands.

Huis in 't Veld, J.W.H., G.J. Monteny and R. Scholtens, 2001. Research into the ammonia emission from livestock production systems no. XLVII: Naturally ventilated cubicle-housing system with a grooved floor for dairy cattle in the summer period (in Dutch). Report IMAG, Wageningen, The Netherlands.

Ivanova-Peneva, S.G., A.J.A. Aarnink and M.W.A. Verstegen, 2006. Ammonia and mineral losses on Dutch organic farms with pregnant sows. Biosystems engineering 93: 221-235.

Kaharabata, S.K., P.H. Schuepp and R.L. Desjardins, 2000. Estimating methane emissions from dairy cattle housed in a barn and feedlot using an atmospheric tracer. Environmental science and technology 34: 3296-3302.

Karlsson, S., 1996. Measures to reduce ammonia losses from storage containers for liquid manure. AgEng 96, Madrid, Spain.

Li, H., H. Xin, Y. Liang, R.S. Gates, E.F. Wheeler and A.J. Heber, 2004. Comparison of direct vs. indirect ventilation rate determination for manure belt laying hen houses. ASAE/CSAE Annual International Meeting, Ottawa, Ontario, Canada.

Monteny, G.J., J.M.G. Hol and J. Mosquera, 2005. Gasvormige emissies. In: Nutriënten management op het melkveebedrijf van de familie Spruit. Studie naar bedrijfsvoering en milieukwaliteit. Report 2005-049. Agrotechnology and Food Innovations, Wageningen, The Netherlands.

Monteny, G.J., J.M.G. Hol, A.C. Wever and R. Scholtens, 1999. Ammonia emission inventory at the regional scale within the nitrogen research programme STOP (in Dutch). Report IMAG, Wageningen, The Netherlands.

Monteny, G.J., M.C.J. Smits, G. Van Duinkerken, H. Mollenhorst and I.J.M. De Boer, 2002. Prediction of ammonia emission from dairy barns using feed characteristics. Part II: Relation between urinary urea concentration and ammonia emission. Journal of dairy science 85: 3389-3394.

Mosquera, J., 2003. Guidelines for the use of passive flux samplers (PFS) to measure ammonia emissions from mechanically ventilated animal houses. Report IMAG, Wageningen, The Netherlands.

Mosquera, J., P. Hofschreuder, J.W. Erisman, E. Mulder, C.E. Van 't Klooster, N.W.M. Ogink, D. Swierstra and N. Verdoes, 2002a. Meetmethoden gasvormige emissies uit de veehouderij. Report 2002-12. IMAG, Wageningen, The Netherlands.

Mosquera, J., P. Hofschreuder and A. Hensen, 2002b. Application of new measurement techniques and strategies to measure ammonia emissions from agricultural activities. Report IMAG, Wageningen, The Netherlands.

Mosquera, J., J.M.G. Hol and J.W.H. Huis in 't Veld, 2004. Onderzoek naar de emissies van een natuurlijk geventileerde potstal voor melkvee. I. A&F Report 324. Agrotechnology and Food Innovations, Wageningen, The Netherlands.

Mosquera, J., G.J. Monteny and J.W. Erisman, 2005. Overview and assessment of techniques to measure ammonia emissions from animal houses: the case of The Netherlands. Environmental pollution 135: 381-388.

Mosquera, J. and N.W.M. Ogink, 2004. Determination of the variation sources associated with ammonia emission measurements of animal housings. AgEng2004, Leuven, Belgium.

Mosquera, J. and N.W.M. Ogink, 2006. New measurement protocol for the determination of NH_3 emission factors from animal houses in The Netherlands. Workshop on agricultural air quality: State of the science, Maryland, USA.

Ni, J.Q. and A.J. Heber, 2001. Sampling and measurement of ammonia concentration at animal facilities. A review. ASAE Annual International Meeting, Sacramento, California, USA.

Ogink, N.W.M. and J.V. Klarenbeek, 1997. Evaluation of a standard sampling method for determination of odour emission from animal housing systems and calibration of the Dutch pig odour unit into standardized odour units. In: International symposium ammonia and odour control from animal production facilities. J.A.M. Voermans and G.J. Monteny, Eds., Vinkeloord, The Netherlands. pp. 231-238.

Pedersen, S., H. Takai, J.O. Johnsen, J.H.M. Metz, P.W.G. Groot Koerkamp, G.H. Uenk, V.R. Phillips, M.R. Holden, R.W. Sneath, J.L. Short, R.P. White, J. Hartung, J. Seedorf, M. Schröder, K.H. Linkert and C.M. Wathes, 1998. A comparison of three balance methods for calculating ventilation rates in livestock buildings. Journal of agricultural engineering research 70: 25-37.

Phillips, V.R., M.R. Holden, R.W. Sneath, J.L. Short, R.P. White, J. Hartung, J. Seedorf, M. Schröder, K.H. Linkert, S. Pedersen, H. Takai, J.O. Johnsen, P.W.G. Groot Koerkamp, G.H. Uenk, J.H.M. Metz and C.M. Wathes, 1998. The development of robust methods for measuring concentrations and emission rates of gaseous and particulate air pollutants in livestock buildings. Journal of agricultural engineering research 70: 11-24.

Phillips, V.R., R. Scholtens, D.S. Lee, J.A. Garland and R.W. Sneath, 2000. A review of methods for measuring emission rates of ammonia from livestock buildings and slurry or manure stores. Part 2: Monitoring flux rates, concentrations and airflow rates. Journal of agricultural engineering research 78: 1-14.

Phillips, V.R., R.W. Sneath, A.G. Williams, S.K. Welch, L.R. Burgess, T.G.M. Demmers and J.L. Short, 1997. Measuring emission rates of ammonia, methane and nitrous oxide from full-sized slurry and manure stores. In: International symposium ammonia and odour control from animal production facilities. J.A.M. Voermans and G.J. Monteny, Eds., Vinkeloord, The Netherlands.

Scharff, H. and A. Hensen, 1999. Methane estimates for two landfills in The Netherlands using mobile TDL measurements. Seventh International Waste Management and Landfill Symposium, Sardinia, Italy.

Scholtens, R., 1993. NH_3 convertor + NO_x analyser (in Dutch). In: Meetmethoden NH_3-emissie uit stallen. Onderzoek inzake mest- en ammoniak- problematiek in de veehouderij. E.N.J. Van Ouwerkerk, Eds. DLO, Wageningen, The Netherlands. pp. 19-22.

Scholtens, R., J.M.G. Hol and V.R. Phillips, 2003a. Improved passive flux samplers for measuring ammonia emissions from animal houses. Part 2: Performance of different types of samplers as a function of angle of incidence of air flow. Journal of agricultural engineering research 85: 227-237.

Scholtens, R., J.M.G. Hol and V.R. Phillips, 2003b. Improved passive flux samplers for measuring ammonia emissions from animal houses. Part I: Basic principles. Journal of agricultural engineering research 85: 95-100.

Secrest, C.D., 2000. Field measurement of air pollutants near swine confined animal feeding operations using UV DOAS and FTIR. In: Water, Ground, and Air Pollution Monitoring and Remediation. T. Vo-Dinh and R.L. Sellicy, Eds. Proceedings SPIE. pp. 98-104.

Sneath, R.W., F. Beline, M.A. Hilhorst and P. Peu, 2006. Monitoring GHG from manure stores on organic and conventional dairy farms. Agriculture, ecosystems and environment 112: 122-128.

Snell, H.G.J., F. Seipelt and H.F.A. Van den Weghe, 2003. Ventilation rates and gaseous emissions from naturally ventilated dairy houses. Biosystems engineering 86: 67-73.

Sommer, S.G., S.M. McGinn, X. Hao and F.J. Larney, 2004. Techniques for measuring gas emissions from a composting stockpile of cattle manure. Atmospheric environment 38: 4643-4652.

Stout, V.L. and T.L. Richard, 2003. Swine methane emissions using the tracer gas ratio method. ASAE Annual international meeting, Las Vegas, Nevada, USA.

Van 't Klooster, C.E. and B.P. Heitlager, 1994. Determination of minimum ventilation rate in pig houses with natural ventilation based on carbon dioxide balance. Journal of agricultural engineering research 57, 279-287.

Van 't Klooster, C.E., B.P. Heitlager and J.P.B.F. Van Gastel, 1992. Measurement systems for emissions of ammonia and other gasses at the Research Institute for Pig Husbandry. Report P3.92. IMAG, Wageningen, The Netherlands.

Vranken, E.C., S., J. Hendriks, P. Darius and D. Berckmans, 2004. Intermittent measurements to determine ammonia emissions from livestock buildings. Biosystems engineering 88: 351-358.

Xin, H.L., H., R.T. Burns, R.S. Gates, D.G. Overhults, J.W. Earnest, L.B. Moody and S.J. Hoff, 2006. Use of CO2 concentrations or CO2 balance to estimate ventilation rate of modern commercial broiler houses. ASAE Annual international meeting, Portland, Oregon, USA.

Zhang, G., J.S. Strøm, B. Li, H.B. Rom, S. Morsing, P. Dahl and C. Wang, 2005. Emission of ammonia and other contaminant gases from naturally ventilated dairy cattle buildings. Biosystems engineering 92: 355-364.

Zhao, Y., W.A. Brewer, W.L. Eberhard and R.J. Alvarez, 2002. Lidar measurement of ammonia concentrations and fluxes in a plume from point source. Journal of atmospheric and oceanic technology 19: 1928-1938.

7. Transitions in agriculture in The Netherlands 1850-2030: towards a healthy and durable production

Oene Oenema

This chapter discusses changes in agriculture, nutrient use efficiency and ammonia emissions in The Netherlands over the last centuries in terms of transitions, and thereby aims at contributing to the debate on the transition towards sustainable agriculture. Firstly, the nature of transitions in agriculture and its driving forces are briefly discussed. Secondly, the transitions in agriculture in The Netherlands are analysed, with emphasis on the last 150 years and an outlook for the next 30 years. Finally, the changes in nutrient use efficiency and ammonia emissions are presented for the period 1800 - 2030.

7.1. Introduction

Agriculture as technology to cultivate land and raise animals was invented in the Near East and China some 10,000 years ago (Balter, 1998; Lev-Yadun *et al.*, 2000; Pringle, 1998). The transformation from a hunting-foraging-gathering lifestyle of the early men to sedentary agriculture with local settlements has been a crucial event in the expansion of the carrying capacity of the land and the subjugation of 'nature'. Progress in traditional agriculture has been achieved steadily over the last 10,000 years, mainly by trial and error, until about 150 years ago when the basis for modern agriculture was developed (Smil, 2000). The transformation from traditional to modern agriculture also marks a crucial event in the expansion of the carrying capacity of the land and the subjugation of nature. Over the past 10,000 years, the carrying capacity increased thousand fold to more than 1000 people per km^2 in some modern farming systems. The major part of the expansion of the carrying capacity was achieved through non-factor inputs, fueled by liberal availability of energy and transport facilities, and by technological developments and improvements in farm management (Smil, 2000, 2001).

With the exception of the last few decades, agricultural developments have been valued as beneficial for humanity. Before 1970 almost nobody thought that farmers' activities threaten nature and the environment. Conversion of forest and natural grassland to agricultural land changed the structure of the landscape and endangered plant and animal species, but it was thought that nature conservation programs could shield biological diversity in protected habitats and national parks. When effects of human activities attained regional and global expressions (Carson, 1962; Hannah *et al.*, 1994; Lubchenco, 1998; Matson *et al.*, 1998; Robertson *et al.*, 2000; Vitousek *et al.*, 1997), the need emerged to reconsider the benefits of agricultural developments. It is generally agreed now that agriculture affects ecosystems at local, regional, continental and even global scales through large-scale cultivation of the land, depletion of scarce resources (e.g. fresh water, biodiversity, fossil energy), and through the emission of ammonia (NH_3) and greenhouse gases (CH_4, N_2O, CO_2) into the atmosphere, and the leaching of nitrate (NO_3^-) phosphorus (P) and chemicals to groundwater and surface waters.

Agriculture of The Netherlands ranks among the highest in the world in terms of production level and resource use per unit surface area (OECD, 2001). Policy makers and farmers largely ignored early signals in the 1960's and 1970's that the intensification of agricultural production during the second half of the 20th century was not sustainable in the long run. However, from 1985 onwards, series of environmental policies and measures have been implemented, especially addressing nitrogen and phosphorus in animal manure, insecticides, fungicides, heavy metals, and the use of land near nature conservation areas. These governmental policies and measures have improved the environmental performance of agriculture, but some problems appear stubborn, and the outbreaks of foot and mouth disease, BSE, pig pest, and Salmonella infections, at the end of the 1990's, have put agriculture further under pressure to reform. New governmental initiatives plead for a 'transition towards sustainable agriculture', by restructuring current agriculture and by having solved all stubborn problems by the year 2030 (Dirven *et al.*, 2002; Rotmans, 2003; VROM, 2001). This governmental plea is in line with various recent studies that indicate that significant systems improvements and efficiency gains in agriculture are needed worldwide in the next decades, to be able to feed the increasing global population and at the same time circumvent large-scale degradation of natural ecosystems and deterioration of ecosystem services through agricultural activities (Smil, 2000; Tilman *et al.*, 2001).

Though there is broad consensus about the need to feed the increasing global population and to circumvent large-scale degradation of natural ecosystems and deterioration of ecosystem services, there is less consensus about 'transitions' and 'transition towards sustainable agriculture'. There are also no 'blueprints' for sustainable agriculture, and no validated 'transition management theory' for sustainable development, which would facilitate managing the transition towards sustainable agriculture (Dirven *et al.*, 2002; Rotmans, 2003). The challenge is to develop sustainable agro ecosystems jointly with all relevant stakeholders, considering all economic, ecological, social and cultural trade-offs of such systems in a balanced manner.

7.2. Transitions in agriculture

7.2.1. Concept and definitions

A transition is defined here as 'a structural change of a (sector or) society, leading to a society (or sector) with fundamental different characteristics'. Examples of transitions include the transition in demography in developing countries (Cipolla, 1978; Rotmans, 2003), the transition of a society using biomass to a society using fossil fuel (oil and natural gas) as extra somatic energy source, the transformation from a hunting-foraging-collecting society to a society with sedentary agriculture and the transformation of traditional farming to modern farming (Dovring, 1965; Smil, 2001). Fundamental to transitions is the change in structure and characteristics of (part of the) society, its large scale, and its large impact. These structural changes occur gradually but continuously in periods of decades to centuries, generally with distinct phases. Large-scale structural changes may include:
1. Changes in number, type, size of agricultural holdings and in the type and total volume of agricultural production.
2. Changes in the relative importance of production factors and resources (land, labour, capital, energy and management).
3. Changes in the organisation and vertical integration of food producing and food processing chains.
4. Changes in ownership of farms and farm land, and in the organisation of farmers and the institutionalisation of farmers' organisations.
5. Changes in total employability and gross domestic production of agriculture.
6. Changes in social integration of the agricultural community within the society.

Evidently, not all changes in agriculture are structural changes, and not all structural changes should be termed 'transitions in agriculture'. Agriculture is highly diverse and complex, and influenced by many external and internal factors. Farmers are often diverse in their responses to natural variability and to changes in for example market, governmental policies and new technology. This makes a discussion about transitions in agriculture subjective (value-oriented) and complex at the same time.

For analysing transitions in agriculture, we used the simple concept depicted in Figure 7.1. External and internal driving forces lead to specific responses by the agricultural community, which transforms the agricultural sector and changes its position in society.

7.2.2. Driving forces and responses

Transitions in agriculture are a response to external and/or internal 'events' that provide the incentive for structural change. Possible events or 'driving forces' for transitions in agriculture are:
1. increasing population pressure;
2. changes in natural conditions (climate, diseases, wars);
3. changes in markets and market prices;
4. innovations and new technology;
5. changes in education, standards, values and judgments;
6. governmental policies and measures.

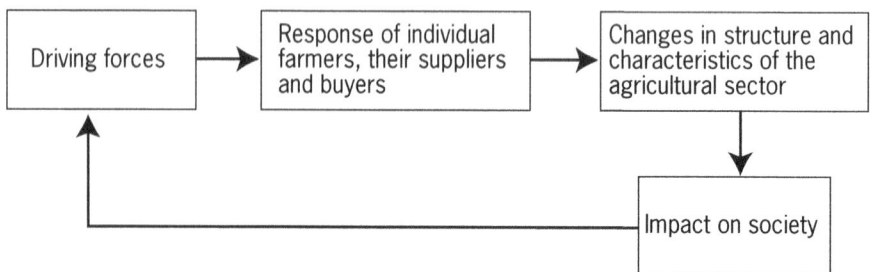

Figure 7.1. General concept of transitions in agriculture. Internal and external factors drive farmers to change, which lead to structural change of the sector, which in turn has impact on the society and may than lead to further change.

In general, the more serious the 'event' and the more attractive the alternative, the greater the response of the agricultural community and the greater the structural change. Responses depend on the education level of the people involved and the adjustability and fragility of the agricultural system. Whether or not the responses lead to 'transitions in agriculture' depends on the collective changes of the whole or greater part of the agricultural community, and not on just a few individual farmers.

7.3. Transitions in agriculture of The Netherlands

The concept of transitions in agriculture outlined above can be used to identify and analyse transitions in agriculture in The Netherlands. Emphasis was on the last 150 years, i.e. the period of 'modern farming', though we realise that the structure of modern farming has its roots in history and as such is the resultant of local environmental conditions and past economic and social developments. We also acknowledge that environmental conditions are regionally highly diverse and that changes in agriculture have been highly diverse regionally (Bieleman, 1987, 1992, 2000; Terpstra, 1980; Van Zanden, 1985), but it is beyond the context of this book to discuss the regional differences in transitions.

Based on the concept of transitions in agriculture discussed before, and on literature and statistical data compiled from annual reports of the Directorate of Agriculture, (Anonymous) we distinguish four transitions in agriculture in the past. The first one is the establishment of traditional agriculture, i.e. the change from nature to farming in harmony with nature, involving also the transition from barter to market economy during the Middle Ages, i.e. prior to 1850. The second one is the transition from traditional to modern farming in the period 1850 - 1950. The third one is the transition from modern to specialised-intensive farming. The fourth one is the transition from specialised-intensive farming to restricted farming, which changed the farmer from entrepreneur to manager dealing with quotas and environmental legislations, in the period 1985 to present. A likely fifth one is the desired transition towards sustainable farming in the period until 2030. The five transitions are briefly described in the following paragraph.

The main driving forces, responses of agriculture to these driving forces, and the subsequent structural changes in agriculture have been summarised in Tables 7.1 and 7.2. Evidently, there is not just one driving force, but a

Table 7.1. Overview of the main driving forces and of the responses by farmers to these driving forces for the transitions in agriculture distinguished between 1850 and 2030 (see text).

	1850 - 1950: from traditional to modern farming	1950 - 1985: from modern to specialised-intensive farming	1985 - present: from specialised-intensive to restricted farming	Present - 2030: towards sustainable farming
Main driving forces	population growth; technology	policies; technology; energy	policies	societal values and standards; policies
Responses of individual farmers	land reclamation intensification	specialisation; intensification; mechanisation; up-scaling	improving management; diversification up-scaling	diversification; up-scaling
Collective response of farmers	co-operatives for supplies, processing & quality control; institutionalisation of interest groups	vertical integration; interest and lobby groups; agribusiness development	agribusiness development; marketing; interest and lobby groups	innovation networks

number of driving forces that drives agriculture to change. Some of these forces have had more influence on a sector or in a region than others, which make it difficult to generalise. Governmental policies and measures appear to be a major driving force for all four transitions. During the first two transitions, governmental policies provided strong incentives for intensification, rationalisation and increasing production. For the third and fourth transitions, governments have exerted strong influence on improving management, diversification (nature management, recreation, etc.) and regulation of production (via quotas). The resulting structural changes in agriculture are manifold, and manifest in various aspects. The development of a strong agribusiness sector, ornamental cultures and the greenhouse sector, all fueled by cheap fossil energy, has given a strong impetus to the economic strength of the agro complex. Currently, three-quarter of the total added value in the whole agricultural sector is generated by the suppliers and processing industry, and only one-quarter by the primary agriculture. The greenhouse sector and the ornamental cultures generate now the most economic revenues, but it is difficult to predict which sector will be most competitive in the next decades. Changes in societal values, standards and judgments are expected to be a main driver for the desired transition towards sustainable agriculture (Table 7.2).

Table 7.2. Overview of the main structural changes in agriculture following the responses of farmers to driving forces, during the four transitions distinguished between 1850 and 2030. See also Table 7.1.

Structural parameter	1850 - 1950	1950 - 1985	1985 - present	Present - 2030
Farm type and structure	mixed systems; peasants with few large owners	specialised; uncoupling of crop & animal production	specialised; vertically integrated	specialised, mixed and high-tech systems
Farm holdings	increased in number, decrease in size	decrease in number, increase in size	strong decrease in number, increase in size	decrease in number, increase in size
Source of income	products; labour hire	products; subsidies	products; subsidies and payments	products; payments for services
Volume of production	doubled	tripled	stabilised	decrease
Main production factors	land; labour; capital	energy; capital; land; labour	management; energy; capital; land	innovation management; energy; capital; land
Vertical integration	many small co-operatives	agribusiness development; contractors	agribusiness development; retailers	production on demand
Horizontal integration	interest groups; sectarianism according to farm size and religion	interest groups; sectarianism according to religion	lobby groups; sectarianism according to farm type	?
Important skill of farmers	craftsman	entrepreneur, craftsman	manager; entrepreneur; craftsman	entrepreneur; networker; steward
Employability (total number)	increased	decreased strongly	decreased strongly	decrease?
Social integration	improved	improved	decreased	improve
Major constraints in agriculture	infrastructure; low income; peasants; low productivity	competitiveness; burden of surpluses	coping with environmental legislation	how to achieve sustainable agriculture?

7.3.1. Pre 1850: from nature to traditional farming in harmony with nature

After the last glacial period, The Netherlands became inhabited some 6,000 years ago, first in the east and south along rivers and streams, but population density remained relatively low (<7 km^{-2}) until the second half of the Middle Ages (McEvedy and Jones, 1978). During the second half of the Middle Ages, the population increased gradually to about 30 km^{-2} in 1500. Thereafter, it increased rapidly further to 70 km^{-2} in 1750 and 100 km^{-2} in 1850. Because of the increasing population, there was an increasing hunger for land (Bieleman, 1992). At the beginning of the Middle Ages more than 80% of the population was working in agriculture (Slicher van Bath, 1960, 1964). At the beginning of the 19th century, still about 50% of the population was working in agriculture, indicating that on average The Netherlands was still an agrarian society at that time (Van Zanden, 1985).

During the Middle Ages, the agriculture society changed from barter to market economy, which allowed specialisation in the society, and which led to a decrease in the percent of the population working in agriculture. There were intricate relationships between population density, agricultural area, livestock density, animal manure and crop yield (Slicher van Bath, 1960). A change in one of these parameters had consequences for one or more of the other. In the course of the Middle Ages, more and more of the natural vegetation (deciduous forest and grasslands) was cleared, so as to feed the growing population. Reclamation of lakes and marshes along the sea into polders extended the surface area by on average 3 to 15 km^2 per year during a period of more than two centuries (Slicher van Bath, 1960) and introduced a polder-like landscape in the western and northern half of the country. The other half of the country was transformed in a highly diverse landscape with half-natural and half-agricultural areas, each with characteristic plant and animal communities. The century-long transfer of plant nutrients via animal manure and litter from common grounds to the arable land around the farm depleted the common grounds and enriched the man-made plaggen soils with nutrients and organic matter, and this contributed further to diversification of the landscape, and to sustaining the production level of the arable land. It is generally believed that agricultural developments before 1850 had a positive influence on landscape diversity and biodiversity. The richness and diversity of plant and animal communities and of landscapes was largest at that time, due to a century-long adaptation of agricultural practices to the local environment, and vice versa (Gorter, 1986).

The western province Holland was already urbanised (>50% of the population living in cities) in the 17th century. It became one of the most densely populated and 'industrialised' areas of Europe, thriving in part on 'trade' with colonies. Life in the cities was unhealthy with death rates exceeding birth rates, while the opposite was true in the country site. Urban waste and dredge materials from canals was shipped to agricultural land around cities to increase productivity. However, productivity was not sufficient to feed the growing urban population; increasing amounts of cereals were imported from abroad from the 17th century onwards.

7.3.2. Period 1850 - 1950: from traditional to modern farming

The industrial revolution, developments in technology and markets, and governmental interference contributed to fundamental changes in agriculture and in the relationship between agriculture, nature and society in the second half of the 19th century and first half of the 20th century. Availability of energy from fossil fuels, and scientific breakthroughs in technology, transport, plant nutrition and crop and animal husbandry increased agricultural and labour productivity and contributed to enlarged markets (Bieleman, 1992; Slicher van Bath, 1960; Smil, 1994; Van Zanden, 1985). These breakthroughs among others led to a flow of cheap crop products from America and Eastern Europe on Western European markets, dropped the prices for agricultural products and led to the agricultural crisis of the 1880's (Dovring, 1965). In response to these developments, the Dutch government for the first time started to systematically stimulate agricultural research and education, and introduced a market and price policy focused on improving the structure and productivity of agriculture and on enlarging the export of agricultural products. The government also stimulated the reclamation of the remaining natural areas and common grounds, mainly heath land on poor soils, drainage of swamps, and the reclamation of new land from sea and lakes. The treaty on land exchange from 1924 facilitated the change from a patchy and highly diversified landscape into a more uniform landscape with larger fields, so as to farm more economically. All theses governmental incentives contributed to increases in agricultural productivity but also to the impoverishment of the rural landscape. Only a few considered these changes as highly negative to nature and humanity and founded the private nature conservation society 'Natuurmonumenten' in 1905, which is now one of the largest society and landowner of The Netherlands.

Farmers started to organise themselves via interest groups and via de establishment of co-operatives for e.g. processing milk into butter and cheese, breeding improved varieties of potatoes, cereals and sugar beet, and breeding improved dairy cattle. This improved the yield, quality and marketing of the products.

Total population in The Netherlands increased exponentially from 3.0 million in 1850 to 10 million in 1950, indicating that population density increased from 100 to 300 inhabitants km^{-2}. The percentage of the labour force employed by agriculture decreased from 50% in 1800 to 44% in 1850 to 20% in 1950. However, the total number of people working in agriculture steadily rose from 0.4 million in 1800 to 0.5 million in 1850 and peaked at 0.75 million in 1947.

The total number of farms steadily increased from about 130,000 in 1800 to 160,000 in 1880 and 235,000 in 1930. Especially the number of small farms increased up to 1930, but decreased thereafter. The increase in number of farms was made possible through reclamation of natural and common areas, and through splitting up the farm property in two or more parts. The area of cereals and cash crops (flax, rapeseed and madder) remained more or less similar, but the area of potatoes, sugar beet and grassland increased. Productivity per unit of land increased as a result of improved management, a change in crops and varieties, and the increased use of fertilisers (e.g. Table 7.3). Yields of crops doubled between 1850 and 1950, but the strongest increased occurred during the second half of this period, coinciding with the increased use of fertilisers (Table 7.3). However, most farmers did not benefit from the increased productivity, and remained poor. This situation was exaggerated by two World Wars and a major economic crisis in between.

7.3.3. Period 1950 – 1985: from modern to specialised-intensive farming

With two World Wars, a serious economic crisis and regular food shortages in mind, there were strong feelings and incentives in European countries after World War II to stimulate the economy and to boost industrial and agricultural productivity. The establishment of the European Economic Community (EEC; later European Union, EU) in 1957 ensued from these general feelings. At that time, Western Europe was a net importer of food, and understandably, the Common Agricultural Policy (CAP) of the EU was strongly focused on stimulating agricultural productivity and stabilising markets. Within a couple of decades, CAP indeed changed the EU from a net importer to a net exporter of food. Initially, these food surpluses were regarded as pleasant, but because of the price support, the surpluses became a financial burden.

The Netherlands is one of the six initial members of the EEC, and its agriculture has benefited from the CAP. The focus on food security and the establishment of common markets within the EU triggered a strong intensification and specialisation of agricultural production. Price support of cereals and butter provided financial security for the farmers and stimulated production. Subsidies provided incentives to farmers for land reclamation and land exchange, for building new animal houses and buying machinery, so as to increase the competitiveness of the agricultural sector. Technological innovations contributed to increases in production and labour productivity. The mechanisation also led to a strong exodus of labour, firstly of contracted labourers, than assisting family and finally of farmers themselves. The quest for labourers by industry and services, and the pressure to work hard for relatively low incomes in agriculture stimulated this exodus.

The rapidly increased productivity and increased production volume surpassed the increasing quest of food by the rapidly growing population during the second half of the 20th century. This led to food surpluses and, because of the low price and income elasticity's for agricultural products on the free market, to decreasing prices for agricultural products (when not supported by the EU). Hence, farmers were in danger to become ruined by their own success (Krielaars, 1965). Decreasing prices and decreasing income provided the incentive to increase production further.

Changes in agriculture were most intense in the period 1960 - 1985 (Table 7.3). In this period, the production of crop and especially animal products increased strongly. The increasing agricultural production was only for a small part for inland consumption. By far the major part (up to 80%) of the agricultural production was exported. The export market appeared to be economically attractive, especially for animal products, flowers and some vegetables.

The large increase in animal production was made possible through the fertiliser induced boost in animal feed production (mainly grass and silage maize) and through the large-scale import of animal feed stuff from outside

Chapter 7

Table 7.3. Changes in some characteristics of agriculture of The Netherlands in the period 1810 to 2000, all data expressed in millions. Changes in total energy use and in the use of natural gas by the whole society are given at the bottom of the table. The energy use by the agro-complex is about 12% of the total use (CBS, 2001; CBS/LEI, 2000; Smaling et al., 1999).

Characteristics, in millions	1810	1850	1880	1910	1920	1930	1940	1950	1960	1970	1980	1990	2000
Surface area agricultural land, ha	1.8	1.9	2.1	2.2	2.2	2.3	2.3	2.3	2.3	2.2	2.1	2.0	2.0
Surface area natural land, ha	0.9	0.8	0.7	0.5	0.5	0.4	0.4	0.3					
Area of glasshouses, ha	0	0	0	0	0.001	0.001	0.003	0.003	0.005	0.007	0.01	0.012	0.016
Milking cows, number	0.7	0.8	0.9	1.1	1.2	1.3	1.5	1.4	1.6	1.9	2.4	1.9	1.5
Pigs, number	0.2	0.2	0.4	1.0	1.5	2.0	1.2	2	2	6	10	14	13
Poultry, number	1.5	1.8	2.5	9.8	9.7	24.6	35	23	43	58	81	91	107
Horses, number	0.2	0.2	0.3	0.3	0.4	0.3	0.3	0.2	0.1	0.05	0.05	0.1	0.2
Tractors, number								0.02	0.08	0.14	0.18	0.18	0.16
Fertiliser N, kg	0	0	1	15	20	53	103	156	224	396	485	412	340
Fertiliser P, kg	0	0	2	24	33	46	47	52	49	48	36	33	27
Fertiliser K, kg	0	0	1	27	48	88	111	128	115	107	93	81	70
Energy use inland economy, PJ								606	925	2014	2732	2702	3024
Use of natural gas, PJ								0	11	635	1274	1290	1467

136

Ammonia, the case of The Netherlands

the EU, which was not regulated by price support. These changes led in part to uncoupling of crop production and animal production; animal production was not limited anymore by the local agricultural surface area and its crop production, but by the import of animal feed from elsewhere. Livestock farmers with little or no land considered animal manure as a waste, instead of a resource. Evidently, the pleasure of food surpluses transformed into the burden of vast surpluses of cereals, milk, and butter, which threatened the economic sustainability of agriculture, and of large surpluses of animal manure which threatened the environmental sustainability.

The specialised-intensive farming became a threat for its own success. The burden of the surpluses changed also the position of the agricultural community in the society. The cost of getting rid of the surpluses of butter and cereals by the EU increased dramatically. The surpluses of animal manure, phosphorus saturated soils, slurry spreaders stacked in mud, and reports about forest dieback due to ammonia volatilisation, and eutrophication of lakes and seas through agricultural nitrogen and phosphorus, all changed the public perception about agriculture. This fueled the establishment of green action groups, which than put increasing pressure on the government to take actions. Farmers were blamed in part for pocketing subsidies and for being environmental criminals (Bloemendaal, 1995; Krielaars, 1965; Lowe and Ward, 1997).

7.3.4. Period 1985 – present: from specialised-intensive farming to restricted farming

Increased awareness of the environmental impacts of the intensification of agricultural production, and of the relationships between the Common Agricultural Policy in the EU and surpluses of butter and cereals, drastically changed the agricultural policies of EU and The Netherlands from the mid 1980's onwards (Brouwer and Lowe, 1998; De Clercq *et al.*, 2001; Henkens and Van Keulen, 2001). Marketable quotas of milk, sugar and starch potatoes, and marketable quotas of animal manure and pigs and poultry were implemented from 1984 onwards and greatly decreased the degrees of freedom for agricultural entrepreneurs. Decreases in the price support of cereals and milk, implementation of set-aside subsidies, and policies and measures on the use of crop protection means and animal manure further restricted farming practices. All these governmental policies were meant to set limits to a further intensification of agricultural production, to decrease non-factor inputs (pesticides, nutrients, energy), and to decrease the emissions of odours, ammonia, nitrogen, phosphorus, heavy metals and chemicals into the wider environment. Codes of good agricultural practices and best management practices were developed and farmers were trained and encouraged to implement these in practice.

Effects of milk quota and of the policies on animal manure and fertilisers are noticeable through decreases in animal density and in use of N and P fertilisers (Table 7.3). The number of dairy cows decreased from 1985 onwards. The number of pigs started to decrease from the end of the 1990's, firstly because of the swine pest and secondly because of the buy-out of pig quotas by the government. The number of poultry continued to increase until poultry quotas were established in 2002, followed by a buy-out of poultry quotas by the government. The implementation (from 1998 onwards) of the nitrogen (N) and phosphorus (P) accounting system MINAS with target N and P surpluses, at farm level, forced farmers to lower the input of N and P via fertilisers, animal feed and animal manure, and to increase the N and P use efficiency. As a result, N and P surpluses in agriculture decreased. More drastic changes in N and P surpluses are expected between 2000 and 2005, when the effects of a step-wise decrease of target N and P surpluses will become noticeable (RIVM, 2002).

In 2005, there were still some 80,000 farm holdings, utilising 2 million ha agricultural land. The dairy sector still used more than 50% of the agricultural land, similar to 150 years ago. Farms with the largest surface area were found in arable farming, while the smallest area per farm was in greenhouse horticulture and floriculture. Over the last decades, the number of farms decreased at a rate of 2 - 3% per year (CBS/LEI, 2000), while the total surface area of agricultural land decreased by 0.3 to 0.5% per year. This indicates that the average size of farms increased steadily. However, this increase in mean farm size, and hence in production volume, was not sufficient to compensate for the loss in income due to increasing costs and lowering prices for agricultural products. As a consequence, more and more farmer families were forced to take additional jobs outside agriculture, to supplement income.

7.3.5. Period present – 2030: from restricted farming towards sustainable agriculture

There is a concern that production quotas and environmental legislations implemented from the mid 1980's onwards will not be sufficient for developing a sustainable agriculture, which meets the targets of tomorrow. The concerns about animal welfare in intensive pig and poultry production, recent outbreaks of animal diseases, high energy use in the intensive horticulture in greenhouses, and the stubborn environmental impacts of current agricultural practices in general, all have contributed to changes in public perception about agriculture. The increasing demands by the society with respect to food quality and safety, animal welfare, landscape and nature conservation, resource use, biodiversity and recreation set increasing constraints to farming systems. In response to the increasing awareness of the unsustainable nature of the intensive agricultural system and the limited effectiveness of environmental policies, a plea was made for a long-term policy view and for structural changes in agriculture through system innovations (VROM, 2001). The year 2030 was chosen as the reference year and sustainability as target, i.e. *'agriculture in harmony with nature and men'* (VROM, 2001). However, sustainability is value-oriented and its objectives can not be laid down as final destination; only 'sustainable directions' can be indicated (Rabbinge, 2000). For calling such directions sustainable, they have to meet the economic, environmental and social-cultural demands by the society at large.

Roughly, three directions have been identified for agricultural development, namely large, broad and smart (Oenema *et al.*, 2006; Vromraad, 2004). Direction 'large' anticipates on the economics of scale, and the economics of specialisation and (to some extent) intensification. Though this direction focuses especially on economic targets, it has the potential of meeting the social-cultural and environmental criteria simultaneously, when strict precautionary measures are taken. Direction 'broad' anticipates on the increasing demands by society to provide social-cultural and environmental services by agriculture. Here, farmers are paid for maintaining the landscape and environmental quality, for providing leisure and recreational areas, for producing biofuels and storing and providing clean surface water, etc. Direction 'smart' anticipates on the prospects of combining technology and ecological principals in smart combination of production processes. This direction requires high-tech, high-skills and large capital investments as well as strong monitoring and control. It may meat economic and ecological targets easily, but the acceptance of this type of 'farms' by the society at large is not without discussion. Prototypes for all three directions are under development, but it is as yet unclear whether these directions really lead to long-term sustainability.

The transition towards a more sustainable agriculture cannot be managed in a traditional way (top-down). Transition management requires facilitating innovations, experimentation, and learning by doing at various levels and by various stakeholders (Oenema *et al.*, 2006; Rotmans, 2003). It's a cyclic and continuous process. This transition management can be organised along food chains, at sector level and at regional level. Directions 'large' and 'smart' may require a food chain approach and/or sector approach, while the direction 'broad' most likely will benefit mostly from a regional approach. Switching from one approach to another can be attractive to reach multiple goals (Oenema *et al.*, 2006).

7.4. Changes in nutrient use and ammonia emissions

The changes in the structure of agriculture coincided (and were made possible in part) with profound changes in the input and output of nutrients, especially nitrogen (N) and phosphorus (P). Mineral fertilisers came on the market from the second half of the 19th century, but use was low become because of lack of knowledge and high prices. Phosphorus (mainly guano) and potassium (mainly kainite) were the dominant fertilisers initially, in part because of their need but also because of their availability. Nitrogen fertiliser became the dominant fertiliser from the 20th century, following the discovery of procedures that allowed producing N fertilisers relatively cheap (Smil, 2001). The use of P and K fertilisers peaked in the 1950s and that of N fertiliser in the 1980s (Table 7.3).

The increased nutrient availability allowed feeding of a larger livestock population. The number of dairy cows steadily increased from about 0.7 million in 1800 to 1.0 million in 1900 and to 2.4 million in 1980. Similarly, the number of pigs and poultry also increased steadily, and between 1950 and 1990 exponentially (Table 7.3 and Figure 7.2).

The strong increase in the number of livestock was made possible through the increased crop yields, but also because of the strongly increasing imports of animal feed from abroad. The amount of P imported via animal feed surpassed the amount of P imported via P fertilisers from the 1970s onwards, and for N from the 1990s onwards (Figure 7.3). Hence, the increased import of animal feed contributed to the increased amounts of N in animal manure (Figure 7.2) and to increased availability of N and P for crop production and increased losses to the environment. Ammonia emission increased steadily

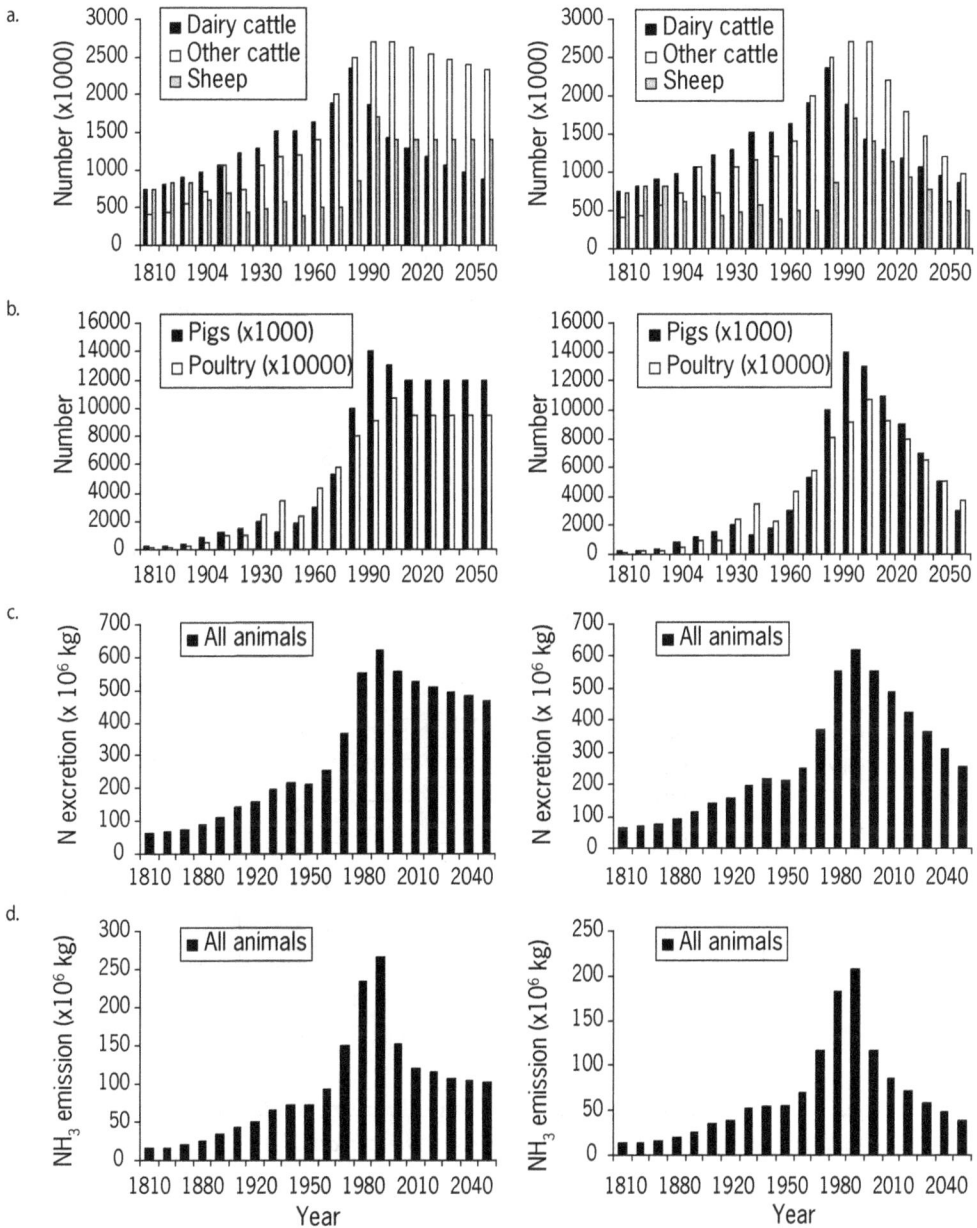

Figure 7.2. Trends in (a) the number of cattle and sheep, (b) the number of pigs and poultry, (c) the total N excretion by all domestic animals, and (d) the total ammonia emissions from animal production systems. Left-hand side trends are based on 'business as usual' scenario, assuming that the milk quota system is in place. Right-hand side panels are based on a strong decrease in other cattle, pigs and poultry so as to fulfill the target of 50 million ammonia emission in 2050.

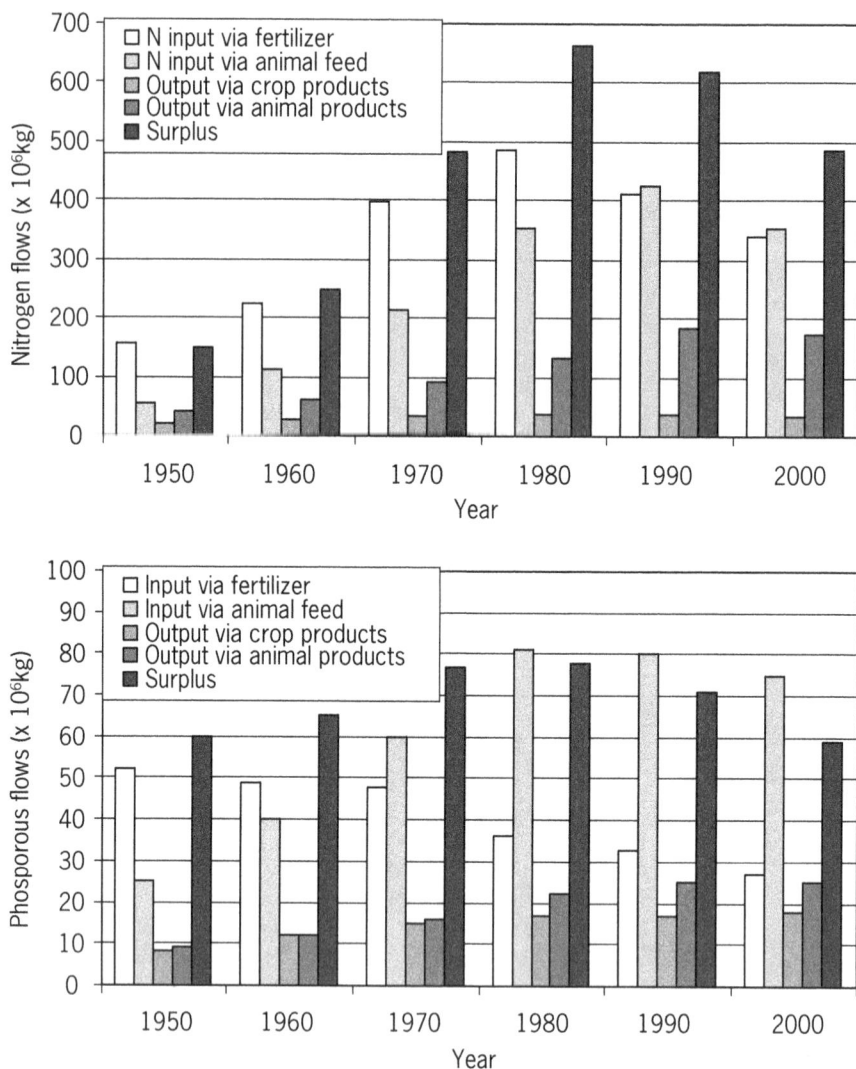

Figure 7.3. Changes in the inputs of nitrogen (upper panel) and phosphorus (lower panel) in agriculture in The Netherlands in the period 1950–2000.

from 1800 onwards, peaked at the end of the 1980s, and decreased rapidly thereafter (Figure 7.2) following the implementation of low emission animal manure application techniques and low emission storage facilities for animal manure described in Chapter 5.

The transitions in agriculture described in the previous paragraphs are evident from the changes in the slope of the curves of crop yield, fertiliser use, animal number and ammonia losses (Figures 7.2 and 7.3). The transition from 1850 to 1950 increased the steepness of the curves shown in these figures only slightly. The transition from 1950 to 1985 led to strong increases in crop yield, animal numbers and ammonia losses, though fertiliser P and K use started to decrease again (Figures 7.2 and 7.3; Table 7.3). The transition from 1985 to present (2005) led to a steady decrease in number of dairy cattle, the use of N fertilisers and the emission of ammonia, while number of pigs and poultry initially increased but afterwards decreased (Figure 7.2). This latter coincided with a decrease in the amount of imported N and P via animal feed (Figure 7.3).

There are no blueprints for the evolution of the number of animals and the use of fertilisers in the transition to sustainable agriculture from present to 2030/2050. However, there are targets for Profit, People and Planet aspects for the transition to sustainable agriculture and there are directions and pathways as indicated in the previous paragraph. Using possible directions and targets, explorations were made of the evolution of animal number, N excretion by livestock and ammonia. Figure 7.2 shows the results of two possible scenarios. The figures on the left-hand side show the results of explorations using the direction 'large', in which the number of dairy cows is determined by the presence of a milk quota system and a steady increase of milk production of 1% per year per cow. The number of other cattle slightly decreases, because of less replacement cattle. The number of pigs and poultry stabilises at 12 and 95 million. The area of agricultural land decreases steadily by 4% per year till 1.6 million ha in 2050, but the area of horticulture and nurseries increases. Emissions of ammonia steadily decrease to about 115 million kg in 2030 and 110 million in 2050.

The figures on the right-hand side of Figure 7.2 show the results of explorations using a combination of the directions 'large' for dairy cattle and 'broad' for other cattle and pigs and poultry. Again, the number of dairy cows is determined by the presence of a milk quota system and a steady increase of milk production of 1% per year. The number of pigs and poultry strongly decrease over time. This scenario leads to a strong decrease in the emissions of ammonia to about 80 million kg in 2030 and 50 million in 2050. Hence, this scenario would allow approaching the long-term target for ammonia emissions in the amount of 50 million kg per year.

The transitions in agriculture have also strong imprints on the N and P use efficiency in agriculture. There are various definitions for nutrient use efficiency (e.g. Mosier *et al.*, 2004), but here we use the output-input ratio for N and P according to the soil surface balance as a measure of the N and P use efficiency. The calculations suggest that the N use efficiency in the pre-fertiliser era was about 0.6 to 0.8 (Figure 7.4). Following the introduction of cheap N fertilisers and the increased import of animal feed, the N use efficiency decreased rapidly to about 0.4 in the 1970s and 1980s. Implementation of policies and measures from the second half of the 1980s onwards on the use of animal manure and fertilisers contributed to slight increases in N use efficiency (from 0.4 to 0.5). Further increases are expected to 0.6 following direction 'large' and to 0.7 following a combination of direction 'large' and 'broad'.

7.5. Discussion and conclusions

Until about 150 years ago, agriculture evolved slowly over millennia, apparently in harmony with nature. Scientific breakthroughs, technological developments and the availability of cheap fossil energy during the last 150 years triggered a cascade of developments, notably in the industrialised countries like The Netherlands, and obstructed this harmony. Various recent reports indicate the large impact of modern agriculture on the

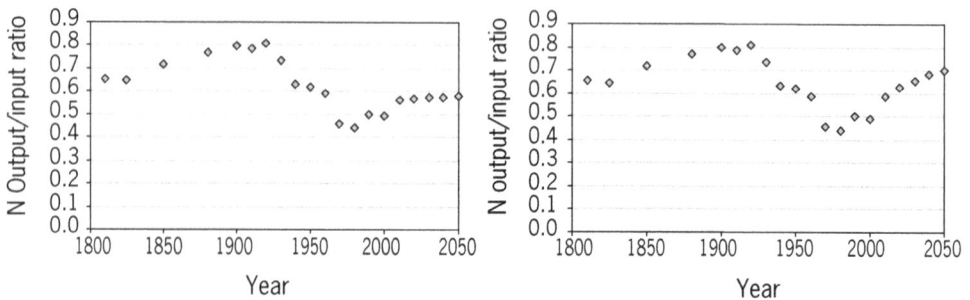

Figure 7.4. Trends in Nitrogen use efficiency in agriculture in The Netherlands in the period 1800–2050. In the left-hand side figure, the trend for the period 2010-2050 is based on 'business as usual' scenario, assuming that the milk quota system is in place. In the right-hand side figure, the trend for the period 2010-2050 is based on a strong decrease in the number of other cattle, pigs and poultry so as to fulfill the target of 50 million kg ammonia emission in 2050.

environment (Tilman *et al.*, 2001; Vitousek *et al.*, 1997). Other studies at the same time indicate that there is ample opportunity for a more productive, environmental less polluting and ecological sound agriculture (Gliessman, 2001; Rabbinge and Van Diepen, 2000; Tilman, 1998), but this challenging outlook still has to be realised.

It remains curious how a small and densely populated country as The Netherlands has become one of the biggest exporters of agricultural products in the world and why nearly 20 years of increasingly tight production quotas and environmental legislations had relatively minor effects on the production levels. Key factors for the boost in agricultural production have been (Bieleman, 2000): (1) the geographical position of The Netherlands in between densely populated areas; (2) the governmental policy strongly supporting a competitive agriculture until 1985; (3) intricate relationships between agricultural research, education and extension; (4) a strong agribusiness, with innovative suppliers and processing industry; (5) abundance of cheap natural gas, supporting fertiliser manufacture and greenhouses; and (6) the foundation of the EU which greatly enlarged the markets. These factors kept agricultural developments going and provide strong economic incentives for intensification of agricultural production. An important question is now how these driving forces interfere in the desired transition towards sustainable agriculture, in which societal, economic and environmental demands and constraints together define agricultural developments.

We defined transitions in agriculture as changes in agriculture that affect its structure and its position in society (reallocation of production factors, type and volume of production, type and size of farms, organisation, vertical integration, employability, and social integration). Because of the large differences in driving forces and the nature of the changes, we identified three transitions during the period 1850 and present. This is unlike other studies (Smil, 2000) which consider only one major transition during the period 1850 - present, i.e. from traditional to modern farming. The fifth one yet to come is the desired transition from restricted farming to sustainable farming between now and 2030.

The transitions have led to strong changes in the use of nutrients and in the loss of N from agriculture, including the emissions of ammonia. Emissions of ammonia peaked in the 1980s and 1990s and thereafter strongly decreased. Decreasing the emissions further to below 50 million kg per year may require a decrease in the number of animals, when following the direction 'broad' (i.e. extensification) or may need the implementation of high-tech solutions when following the direction 'smart'. Realising the target of 50 million kg of ammonia emission per year would bring the total emissions back to the level that occurred at the beginning of the 20th century, before the start of the large-scale use of N fertilisers and the strong increase of the number of animals during the second half of the 20th century. Indeed, realising this target will be a major achievement of Dutch agriculture.

References

Anonymous, Verslagen en mededelingen van den directie van den landbouw. Report Directie Landbouw, The Hague, The Netherlands.

Balter, M., 1998. Why settle down? The mystery of communities. Science 282: 1442-1445.

Bieleman, J., 1987. Farming on sandy soil in Drente 1600-1910. A new view on traditional farming. A.A.G. Bijdragen 29.

Bieleman, J., 1992. History of agriculture in The Netherlands 1500-1950. Changes in diversity (in Dutch). Boom Meppel, Amsterdam, The Netherlands.

Bieleman, J., 2000. Agriculture. In: Technique in The Netherlands in the twentieth century, Volume III agriculture and food. Walburg Pers, Zutphen, The Netherlands.

Bloemendaal, F., 1995. Swamp of manure (in Dutch). SDU publishers, The Hague, The Netherlands.

Brouwer, F.M. and P. Lowe, 1998. CAP and the rural environment in transition. A panorama of national perspectives. Wageningen Pers, Wageningen, The Netherlands.

Carson, R.L., 1962. Silent spring. Houghton Mifflin, Boston.

CBS, 2001. CBS Statline. Data availableat http://www.cbs.nl/.

CBS/LEI, 2000. Agriculture, environment and economy 1999 (in Dutch). Landbouw-Economisch Instituut LEI, The Hague, The Netherlands.

Cipolla, C.M., 1978. The economic history of world population. Penguin Books, London, England.

De Clercq, P., A.C. Gertsis, G. Hofman, S.C. Jarvis, J.J. Neeteson and F. Sinabell, 2001. Nutrient management legislation in European countries. Wageningen Press, Wageningen, The Netherlands.

Dirven, J.M.C., J. Rotmans and A.P. Verkaik, 2002. Society in transition; an innovative view. Ministry of LNV, The Hague, The Netherlands.

Dovring, F., 1965. The transformation of European agriculture. Cambridge economic history of Europe, Cambridge, England.

Gliessman, S.R., 2001. Agroecosystem sustainability. Developing practical strategies. CRC Press LLC, Florida, USA.

Gorter, H.P., 1986. Room for nature, 80 years working on nature in the future (In dutch). Natuurmonumenten, 's Gravenland, The Netherlands.

Hannah, L., D. Lohse, C. Hutchinson, J. Carr and A. Lankerani, 1994. A preliminary Inventory of human disturbance of world ecosystems. Ambio 23: 246-250.

Henkens, P.L.C.M. and H. Van Keulen, 2001. Mineral policy in The Netherlands and nitrate policy within the European Community. Netherlands journal of agricultural science 49: 117-134.

Krielaars, F.W.J., 1965. Agricultural problems emerging with economic growth. Possibilities and constraints of governmental policies and measures in affluent countries (In Dutch). Stenfert Kroese NV, Leiden, The Netherlands.

Lev-Yadun, S., A. Gopher and S. Abbo, 2000. The cradle of agriculture. Science 288: 1602-1603.

Lowe, P. and N. Ward, 1997. The moral authority of regulation: the case of agricultural pollution. In: Controlling mineral emissions in European agriculture. Economics, policies and the environment. E. Romstad, J. Simonsen and A. Vatn, Eds. CAB International, Wallingford, England. p. 59-72.

Lubchenco, J., 1998. Entering the century of the environment: a new social contract for science. Science 279: 491-497.

Matson, P.A., R. Naylor. and I. Ortiz-Monasterio, 1998. Integration of environmental, agronomic and economic aspects of fertiliser management. Science 280: 112-115.

McEvedy, C. and R. Jones, 1978. Atlas of world population history. Harmondsworth.

OECD, 2001. Environmental indicators for agriculture: Methods and results. OECD, Paris, France.

Oenema, O., J.W.H. Van der Kolk and A.M.E. Groot, 2006. Agriculture and environment in transition. WOT Report. Alterra, Wageningen, The Netherlands.

Pringle, H., 1998. The slow birth of agriculture. Science 282: 1446-1450.

Rabbinge, R., 2000. Sustainability and sustainable development (In Dutch). Wageningen University, Wageningen, The Netherlands.

Rabbinge, R. and C.A. Van Diepen, 2000. Changes in agriculture and land use in Europe. European journal of agronomy 13: 85-99.

RIVM, 2002. MINAS and environment, balance and exploration (in Dutch). RIVM, Bilthoven, The Netherlands.

Robertson, C.P., E.A. Paul and R.R. Harwood, 2000. Greenhouse gases in intensive agriculture: contribution of individual gases to the radiative forcing. Science 289: 1922-1925.

Rotmans, J., 2003. Transititiemanagement: Sleutel voor een duurzame samenleving. Koninklijke van Gorcum, Assen, The Netherlands.

Slicher van Bath, B.H., 1960. The agricultural history of western Europe 500-1850 (In Dutch). Het Spectrum, Utrecht, The Netherlands.

Slicher van Bath, B.H., 1964. Eighteen century agriculture on the continent of Europe: Evolution or revolution. Agricultural history 43: 164-180.

Smaling, E.M.A., O. Oenema and L.O. Fresco, 1999. Nutrient disequilibria in agroecosystems: Concepts and case-studies. CAB International, Wallingford, England.

Smil, V., 1994. Energy in world history. Westview, Boulder, USA.

Smil, V., 2000. Feeding the world. A challenge for the twenty-first century. MIT Press, Cambridge, Massachusetts, USA.

Smil, V., 2001. Enriching the earth. Fritz Haber, Carl Bosch and the transition of world food production. MIT Press, Cambridge, Massachusetts, USA.

Terpstra, P., 1980. Hundred years of Frisian agriculture (In Dutch). Van Seijen, Leeuwarden, The Netherlands.

Tilman, D., 1998. The greening of the green revolution. Nature 396: 211-212.

Tilman, D., J. Fargione, B. Wolff, C. D'Antonio, A. Dobson, R.W. Howarth, D. Schindler, W.H. Schlesinger, D. Simberloff and S. D., 2001. Forecasting agriculturally driven global environmental change. Science 292: 281-284.

Van Zanden, J.L., 1985. The economic development of agriculture in The Netherlands 1: 800-1914. Hes Studia Historica Publishers, Utrecht, The Netherlands.

Vitousek, P.M., H.A. Mooney, J. Lubchenko and J.M. Melillo, 1997. Human domination of earth's ecosystems. Science 277: 494-499.

VROM, 2001. A world and a will; Towards sustainability (in Dutch). Nationaal milieubeleidsplan 4.

Vromraad, 2004. Meerwerk. Advies over de landbouw en het landelijk gebied in ruimtelijk perspectief. Advies 042, The Hague, The Netherlands.

Appendices

Appendix 1. Inventory of ammonia emission measurements on cattle houses

The ammonia emission from different housing systems as measured by the ASG gaseous emissions team are given in Tables 1 and 2. In Table 1 results in dairy cow housing systems are given per season because in some cases no results in summer were available. The listed measurements were performed to evaluate the emission level of systems that applied to be taken up into the Rav list. In the past in dairy cattle only the ammonia emission during the winter housing period (190 days) was evaluated. From 2002 onwards also the summer season (175 days) is included. All measurements presented in Table 1 were performed before the governmental list of emission factors was updated in 2002. The emission factors given in Chapter 5.1.1 in Table 5.1 are based on the update in 2002. In the update it was taken into account that nitrogen utilisation was improved in recent years: from the start of the routinely determination of bulk milk urea concentrations in The Netherlands in 1998, until 2002 the average urea content was decreased from 30 to 25 mg/100 ml milk.

In the Kamerik case (cubicles & slatted floor) in 1996 high emission rates were measured during the first months. This could have been due to the newly installed concrete slatted floor that could have resulted in a high pH at the concrete surface where the urine puddles were emitting. In the Achterberg case (tying stall) forced ventilation rates and animal production levels fell below the average values in The Netherlands.

Table 1. Emission results from dairy cow houses.

Ammonia emission per animal place (kg NH₃ per season)

Location	1st period	2nd period	Ref[1]	Season[2]	Year	System description
Arriën	5.5			S	1998	cubicles, solid floor with
Arriën		4.21	4.4	W	1998	parallel grooves and perforations
Kamerik	8.9		8.8	W	1996	cubicles, concrete slatted
Kamerik		7.0		S	1996	floor
Kraanmeer	4.0		4.4	W	1995	cubicles, inclined solid floor and spraying scrapers
Oosterhesselen	6.8		8.8	W	1993	deep litter housing
Achterberg	2.0			W	1991	tying stall

[1]Ref: reference emission level for this system in winter season (190 days). The threshold level of emission reducing systems was in the past set at 50% of the level of the traditional system.
[2]Emissions are expressed per season: S = summer 175 days including grazing hours; W = winter 190 days without grazing.

Appendix 1. Inventory of ammonia emission measurements on cattle houses

Table 2. Data on veal calves (kg ammonia per animal place per year).

Location	1st period	2nd period	System description
Kootwijkerbroek	2.7	2.4	group housing wooden slatted floor above pit
Kootwijkerbroek	3.0	2.7	group housing plastic slatted floor above pit
Kootwijkerbroek	2.4	2.0	group housing wooden floor above flushing gutters
Ede		2.7	group housing wooden slatted floor above pit

In veal calves houses emissions were simultaneously compared of three group housing systems units within one farm in Kootwijkerbroek in 1997.

The group housing system with a wooden slatted floor above the pit is the most common in The Netherlands. In 2004 the emission level of one unit in Ede with the same characteristics was measured to evaluate the earlier measured emission level. The results were approximately 10% higher than the official emission factor of 2.5 kg ammonia per animal place per year (Table 5.1, Chapter 5.1.1) that was already incorporated in the law. While nitrogen utilisation is decreasing and nitrogen intake is increasing with age, veal calves that are kept for more than half a year will result in higher average emission rates.

In Table 3 emission rates in several situations are given based on more recent measurements in a cubicle house with a slatted floor (Van Duinkerken *et al.*, 2003, 2005).

Table 3. Ammonia emission from a cubicle cow house with a slatted floor, depending on season, milk urea content and grazing system (Monteny et al., 2001).

	Ammonia emission (kg ammonia per cow)[1]								
	Winter season			Summer season			Total per year		
Milk urea concentration (mg urea/100 g milk)	20	30	40	20	30	40	20	30	40
Grazing system									
Grazing day & night	4.5	6.2	7.9	*2.7*	*3.8*	*4.8*	*7.2*	*10.0*	*12.7*
Restricted grazing	4.5	6.2	7.9	*4.1*	*5.5*	*7.0*	*8.6*	*11.7*	*14.9*
Summer feeding, zero-grazing	**4.5***	**6.2**	**7.9**	**5.3**	**7.3**	**9.3**	9.8	13.5	17.2

[1]Results in bold were actually measured while summer feeding (grass silage and/or maize silage) was applied (Van Duinkerken *et al.*, 2003; Van Duinkerken *et al.*, 2005). Results in italic were calculated assuming that the cows stay indoors only 4 hours, around milking, when grazing day & night and assuming that the cows stay indoors 14 hours per day, when restricted grazing. Per grazing hour the calculated emission reduction from the cow house is 2.4%.

Further references

Monteny, G.J., J. Huis in 't Veld, G. Van Duinkerken, G. André and F. Van der Schans, 2001. Towards ammonia emission factors per year for dairy cow housing systems. IMAG Report 2001-09. IMAG, PV and CLM, Wageningen, The Netherlands. (in Dutch)

Van Duinkerken, G., G. André, M.C.J. Smits, G.J. Monteny, K. Blanken, M.J.M. Wagemans and L.B.J. Sebek, 2003. Relationship between diet and ammonia emissions from a dairy cow house (in Dutch). PV Report 25. PV / IMAG, Wageningen, The Netherlands.

Van Duinkerken, G., G. André, M.C.J. Smits, G.J. Monteny and L.B.J. Sebek, 2005. Effect of rumen-degradable protein balance and forage type on bulk milk urea concentration and emission of ammonia from dairy cow houses. Journal of dairy science 88: 1099-1112.

Appendix 2. Tie stall (tied cow housing)

System description

In tied cow housing with slurry, a limited slurry pit area is covered with a slatted floor with iron grids. Compared to loose housing systems, the emitting floor area in these housing systems is small because tying the cow restricts the spread of the cow excreta over a limited floor area. In addition, the emitting area of the slurry storage behind the cow can be small.

Though frequently used in the past, this housing system nowadays is used less than 10% and has hardly been built in recent decades. Milking can be done in a modern separate milking parlour or at the individual stall (with tubes for the vacuum to be connected to the milking machine units and optionally also a milk transporting tube).

Remarks

Because animal welfare is impaired by tying the cow during the winter season, this system is not recommended. Farm management on this type of farm can be qualified as 'old-fashioned'.

Key parameters

Emitting area is small due to the tying of the cows and the small emitting area of the slurry storage behind the cow (maximum 1 m^2/cow). Storage time of slurry was restricted: 10-14 days. Also a lower level of (forced) ventilation compared to naturally ventilated loose houses may play a part in the lower emission.

Key parameters during emission measurements

The system has forced ventilation with an average of 100 m^3/hr per cow (within the 65-115 m^3/hr interval). The diet during winter: grass (*ad liberty*), beet pulp (6 kg), maize silage (3 kg) and concentrates (3 kg). The diet in spring: fresh grass, maize silage (8 kg) and standard concentrates (4.5 kg).

* Body weight: 550 kg (breed: MRY).
* Production level: 6,500 kg milk per year.
* Season: winter and spring.

Table 1. Housing and grazing periods and average temperatures.

Situation	from	until	T_{indoor}
Total period	22/12	21/5	
Cows permanently inside, doors closed	22/12	6/3	15.4
Cows permanently inside, doors open during the day if T is high	6/3	25/4	19.2
Cows not indoors between 9.00 and 16.45[a]	25/4	21/5	17.3

[a]For the purpose of the measurements cows were kept indoors during the night in this period.

Ammonia, the case of The Netherlands

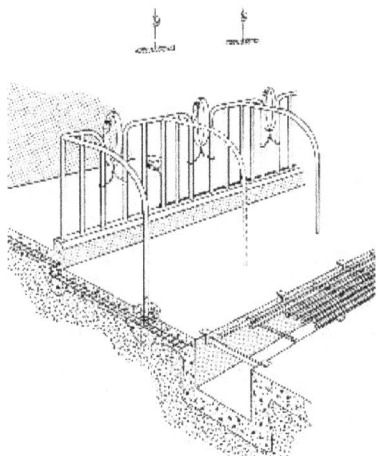

Emission reduction using this system

The emission data during winter lead to an emission level of 2 kg per animal in 190 days. Emission factors given in the Rav list are 4.3 kg per animal place per year (365 days, including summer with grazing in summer season). Therefore, a reduction of 55% was reached, as compared to a traditional loose house with cubicles, a slatted floor and with grazing in summer.

Extrapolation to similar systems and animal categories

The Rav list emission factor only refers to tied housing with slurry and a maximum slurry pit area of 1.2 m² per cow, covered with a slatted floor with iron grids. Part of the rest of the building may also be used for slurry storage. Such a storage facility should not be in direct connection with the above floor compartment (i.e. be air tight covered by a solid concrete floor) and it should be constructed according to Dutch directives for slurry storages (Frénay, 1993).

The emission factor is applicable to both naturally ventilated and forced ventilated tied cows housings. The emission factor is only applicable to grazing mature dairy cows (incl. heifers); it is not applicable to young livestock and it is not applicable to zero grazing cows.

Further references

Frénay, L., 1993. Handleiding bouwtechnische richtlijnen mestbassins (HBRM), CUR / IMAG-DLO, Gouda / Wageningen, The Netherlands, rapport 91-10 / 91-13, 115 pp.

Groenestein, C.M. and H. Montsma, 1991. Field research into the ammonia emission from animal housing systems: Tying stall for dairy cattle (In Dutch). Report 91–1002, Agricultural Research Department, Wageningen, The Netherlands, 14 pp. (excl. appendices).

Infomil. http://www.infomil.nl/

Monteny, G.J., J. Huis in 't Veld, G. Van Duinkerken, G. André en F. Van der Schans, 2001. Naar een jaarrond emissie van ammoniak uit melkveestallen. Gezamenlijk rapport IMAG, PV en CLM. IMAG rapport 2001-09, Wageningen, The Netherlands, 27 pp.

Appendix 3. Solid sloping floor with scraper and urine gutter

System description

The cattle housing with a solid sloping floor, scraper and central urine gutter is known in the Dutch Rav list under A.1.3. Manure is removed from the solid floor every 2 hours with a 3 m wide scraper that is equipped with a rubber strip. The scraper can be pulled by a cable or chain that is driven by an electric engine, placed in or above the central urine gutter. The emission reduction is obtained in two ways. Part of the drained urine is collected in the urine gutter, potentially reducing ammonia production on the floor surface. Secondly, the slurry pit is closed by both the solid floor and rubber slabs at the floor ends, where the faeces collected by a scraper are dumped into the pit. The closure of the slurry pit is expected to decrease both the air velocity above the slurry surface and the exchange of air between the house and the pit, thus decreasing the escape of ammonia produced in the pit. Airflow over the slurry in the pit, induced by temperature differences between the outside and the inside air, are considered the driving force for emissions from the pit. Pit emission is only eliminated with certainty if all air exchange between pit and house is prevented.

Ventilation system

Most loose houses in NL are naturally ventilated and the emission factor also was based on the reduction in a naturally ventilated cow house. The system is expected to be adequate in forced ventilated cow houses as well.

Key parameters

The maximum total walking floor area per cow should be less or equal to 3 m^2 (including the cow walking alley, passages and the waiting room near the milking parlour). A spraying system should be installed on the floor in order to prevent serious slipperiness. Typically, the floor is wetted with a fluid with low ammonia content like rinsing water from the milking machine or fresh water. Preventing air exchange with the under-floor slurry pit is crucial in this system. To facilitate a quick removal of urine, the slope of the floor should be 2-3% towards the urine gutter in the centre. The concrete floor surface should be produced and finished according to Dutch directives (CUR recommendation 57).

Operational control for safeguarding emission reductions

- Proper maintenance and functioning of the scraper and its rubber or synthetic strip.
- Slurry discharge openings at the end of the alleys should be kept well enclosed to prevent air exchange between the slurry pit and the cow house.
- Mixing should be done regularly to prevent heaps of solid manure to build up below the discharge opening in the slurry pit.
- During mixing the slurry special care should be taken for exchange of accumulating poisonous gases in the head space above the slurry; maximum ventilation during mixing is recommended at all times, and especially when mixing is not done regularly.

A new solid sloping floor (V shape) with a urine gutter being installed in the middle.

Key parameters during emission measurements

Several variants of this floor were tested. The diets varied. Grass (fresh or silage), maize silage and concentrates were the main parts of the diet.
- Body weight: 600 kg (breed: HF).
- Production level: approx. 7,000 kg milk per year.
- Season: autumn/winter/spring/summer.

Emission reduction using this system

With grazing (at least part of the day) in summer: 7.5 kg per animal place per year (365 days). Without grazing in summer: 8.6 kg per animal place per year (365 days). Therefore a reduction of 21% and 22% was calculated respectively when comparing these data to values obtained from a traditional loose house with cubicles and a slatted floor.

Extrapolation to similar systems and animal categories

Several types of scrapers and floor wetting can be used. Floors may be made up of prefabricated elements (joints between elements should be air tight) or the floor could be made *in situ*.

Further references

Monteny, G.J., J. Huis in 't Veld, G. Van Duinkerken, G. André and F. Van der Schans, 2001. Naar een jaarrond emissie van ammoniak uit melkveestallen. Gezamenlijk rapport IMAG, PV en CLM. IMAG rapport 2001-09, Wageningen, The Netherlands, 27p.

Scholtens, R., J.J.C. Van der Heiden-de Vos and J.W.H. Huis in 't Veld, 1996. Praktijkonderzoek naar de ammoniakemissie van stallen XXX: natuurlijk geventileerde ligboxenstal voor melkvee met hellende dichte vloer en zelfrijdende sproeischuiven. DLO-rapport 96-1006, Wageningen, The Netherlands, 20 pp.

Scholtens, R. and J.W.H. Huis in 't Veld, 1997. Praktijkonderzoek naar de ammoniakemissie van stallen XXXVI: natuurlijk geventileerde ligboxenstal met betonroosters voor melkvee. DLO-rapport 97-1006, Wageningen, The Netherlands, 35 pp.

Appendix 4. Cow cubicle cow house with grooved floor

System description

The grooved floor system consists of solid levelled prefabricated concrete elements with grooves covering a slurry pit. The faeces were removed every two hours by a mechanical scraper and were dumped into the pit through a floor opening at the end of the alley. The alley floors should provide a stable surface for the animals to walk on. The surface texture of the floor surface is slip resistant to support the animals' movement. The urine could drain along the grooves. Perforations in the grooves were spaced 1.1 m apart. Urine could be drained trough the perforations directly into a slurry pit below. The total area of perforations of the floor elements was less than 0.5% of the total floor area. The system is known in the Dutch Rav list under A.1.5.

Ventilation system

Most loose houses in NL are naturally ventilated and the emission factor also was based on the reduction in a naturally ventilated cow house. The system is expected to be adequate in forced ventilated cow houses as well.

Key parameters

Essential features in operating this system are:
- The proper functioning of the scraper and its rubber or synthetic blade (good maintenance).
- Urine should run trough the perforations into the pit as quickly as possible. The herdsman should undo the possibly clogging with solid manure or feed of the perforations in the floor.
- Slurry discharge openings at the end of the alleys should be kept well enclosed to prevent air exchange between the slurry pit and the cow house.
- Mixing should be done daily to prevent heaps of solid manure to build up below the discharge opening in the slurry pit.
- During mixing the slurry special care should be taken for exchange of accumulating poisonous gases in the head space above the slurry; maximum ventilation during mixing and while opening the pit is recommended especially when mixing is not done at the recommended daily frequency.

Key parameters during emission measurements

Diets differed between seasons and between the two farms. Grass (fresh or silage), maize silage and concentrates were the main parts of the diet. Some additional low protein products were also included.
- Body weight: 600 kg (breed: HF).
- Production level: 7,500-8,000 kg milk per year.
- Season: autumn/winter/spring/summer.

During summer restricted grazing (siesta system: a limited number of hours grazing after milking) was applied. This may result in a better animal utilisation of nitrogen. Farm management was above average and nitrogen excretion was probably below average.

Emission reduction using this system

With grazing (at least part of the day) in summer: 7.7 kg per animal place per year (365 days). Without grazing in summer: 9.2 kg per animal place per year (365 days). Therefore a reduction of 19% and 16% was calculated respectively when comparing these data to values obtained from a traditional loose house with cubicles and a slatted floor. The emission factors are based on measurements in one commercial farm (Huis in 't Veld and Scholtens, 1998; Huis in 't Veld *et al.*, 2001) and in experimental farm 'De Marke' (Monteny *et al.*, 2001).

Extrapolation to similar systems and animal categories

The emission factor is applicable to naturally ventilated cow housings. In forced ventilated cow housings reductions (expressed as a proportion of a forced ventilated house with a conventional slatted floor) may be of the same extent.

The emission factor is only applicable to mature dairy cows (incl. heifers); it is not applicable to young stock. For young stock an analogous floor with different dimensions has been developed, taking into account the sizes of the claws of calves. The emission factor of this system has not yet been measured. The floor should be non-sloping.

Further references

Huis in 't Veld, J.W.H. and R. Scholtens, 1998. Praktijkonderzoek naar de ammoniakemissie van stallen XXXXII: Natuurlijk geventileerde ligboxenstal met sleufvloer voor melkvee. Dienst Landbouwkundig Onderzoek Rapport 98–1006. Wageningen, The Netherlands.

Huis in 't Veld, J.W.H., G.J. Monteny and R. Scholtens, 2001. Praktijkonderzoek naar de ammoniakemissie van stallen XLVIII: Natuurlijk geventileerde ligboxenstal met sleufvloer voor melkvee; zomerperiode. IMAG rapport 2001-03. Wageningen, The Netherlands.

Monteny, G.J., J. Huis in 't Veld, G. Van Duinkerken, G. André and F. Van der Schans, 2001. Naar een jaarrond emissie van ammoniak uit melkveestallen. Gezamenlijk rapport IMAG, PV en CLM. IMAG rapport 2001-09, Wageningen, The Netherlands, 27 pp.

Swierstra, D., 1998. Stalvloer alsmede vloerelement voor toepassing in een dergelijke vloer. Patent No. NL1003271.

Swierstra, D., C.R. Braam and M.C.J. Smits, 2001. Grooved floor system for cattle housing: ammonia emission reduction and good slip resistance. Applied engineering in agriculture 17: 85-90.

Appendix 5. Regular housing system

System description
Regular pig housing systems in The Netherlands have the following characteristics:
- Bare concrete partly (sows, fatteners, weaners) of fully (weaners) slatted floors.
- Mechanically ventilation (all pig houses) and heating systems (especially in farrowing and weaning rooms). Floor heating is popular.
- Manure storage underneath the slatted floor for 1 to 6 months.
- 10 to 20 pigs (weaners, fatteners) pigs per pen.
- Individual housing in crates or group housing for dry and pregnant sows.

Key parameters during emission measurements for fattening pigs
The measuring location should (preferably) fulfil the following demands:
- The animal house should be in use for at least one fattening period.
- The animals should be kept according to the most recent legislation on animal welfare.
- The number of pigs in the pen should be between 10 and 40.
- The CO_2 concentration inside the room fall below the maximum of 3,000 ppm.
- The protein content of the diet is minimal 155 g/EW for starting feed and minimal 150 g/EW for finishing feed.
- Water should be available *ad libitum.*
- The minimal growth of fatteners from 25-115 kg should be 750 g/d.
- The number of culled pigs is maximal 5%.
- A minimum number of 50 pigs in the room.

The following data should be registered by the farmer during the measurements:
- The total feed given per room and fattening period.
- The number of animals inside the room during the whole fattening period.
- The dates of manure removal from the manure pit.
- A list of veterinary treatments at group level.

As a check the CO_2-concentration in each room, on each farm, and in each measuring period is measured with Kitagawa tubes.

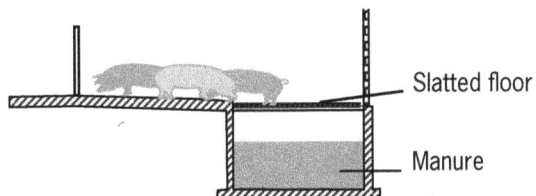

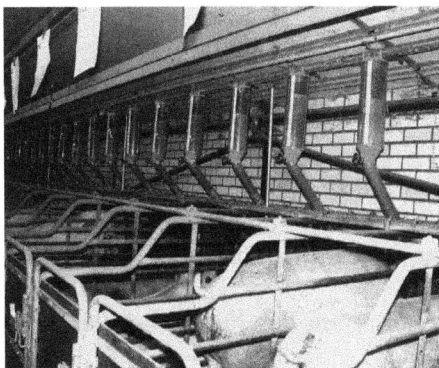

Emission reduction using this system

Table 1. Emission factors for traditional systems in pig production (kg NH₃/y per pig place).

Weaners	Farrowing sows, including piglets	Dry and pregnant sows	Fattening pigs	Free range pigs
0.60	8.3	4.2	2.5	3.5

Further references

VROM, 2006. Regeling ammoniak en veehouderij. Staatscourant 207 (October).

System description

Ammonia emission in this system is reduced by mechanical removal of the fresh manure in the pit using an automated scraper or a belt system.

Key parameters

The main critical success factor of this system is that the manure should be removed completely, otherwise the emitting area will still be present.

Key parameters during emission measurements for fattening pigs

The measuring location should (preferably) fulfil the following demands:
- The animal house should be in use for at least one fattening period.
- The animals should be kept according to the most recent legislation on animal welfare.
- The number of pigs in the pen should be between 10 and 40.
- The CO_2 concentration inside the room fall below the maximum of 3,000 ppm.
- The protein content of the diet is minimal 155 g/EW for starting feed and minimal 150 g/EW for finishing feed.
- Water should be available *ad libitum*.
- The minimal growth of fatteners from 25-115 kg should be 750 g/d.
- The number of culled pigs is maximal 5%.
- A minimum number of 50 pigs in the room.

The following data should be registered by the farmer during the measurements:
- The total feed given per room and fattening period.
- The number of animals inside the room during the whole fattening period.
- The dates of manure removal from the manure pit.
- A list of veterinary treatments at group level.

As a check the CO_2-concentration in each room, on each farm, and in each measuring period is measured with Kitagawa tubes.

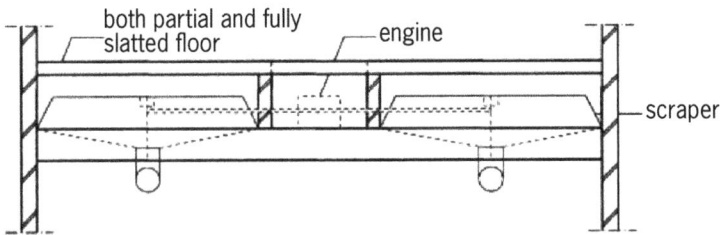

Emission reduction using this system

Table 1. Emission factors for the system with regular complete removal of the manure using a mechanical scraper (kg NH₃/y per pig place).

Weaners	Farrowing sows, including piglets	Dry and pregnant sows	Fattening pigs
0.18 – 0.20	2.5 – 4.0	2.2	-

Further references

VROM, 2006. Regeling ammoniak en veehouderij. Staatscourant 207 (October).
Infomil. http://www.infomil.nl/contents/pages/22613/bb_93-03-001_v1.pdf

Appendix 7. Flushing manure with acidified liquid manure

System description
Ammonia emission in this system is reduced by collecting fresh urine and faeces in an acidified liquid. This mixture is regularly removed and replaced by new acidified liquid. The liquid is obtained by separating the faeces particles from the liquid part, followed by an acidification step to a pH below 6. The layer of fresh acidified liquid should be at least 5 to 10 cm high and the flushing should be done with a frequency of at least once every two days.

Key parameters
The main critical success factor is that the acidified liquid in the manure pit is regularly replaced by new liquid. The pH of the liquid at the time of removal should not be higher than 6.5.

Key parameters during emission measurements for fattening pigs
The measuring location should (preferably) fulfil the following demands:
- The animal house should be in use for at least one fattening period.
- The animals should be kept according to the most recent legislation on animal welfare.
- The number of pigs in the pen should be between 10 and 40.
- The CO2 concentration inside the room fall below the maximum of 3,000 ppm.
- The protein content of the diet is minimal 155 g/EW for starting feed and minimal 150 g/EW for finishing feed.
- Water should be available ad libitum.
- The minimal growth of fatteners from 25-115 kg should be 750 g/d.
- The number of culled pigs is maximal 5%.
- A minimum number of 50 pigs in the room.

The following data should be registered by the farmer during the measurements:
- The total feed given per room and fattening period.
- The number of animals inside the room during the whole fattening period.
- The dates of manure removal from the manure pit.
- A list of veterinary treatments at group level.

As a check the CO_2-concentration in each room, on each farm, and in each measuring period is measured with Kitagawa tubes.

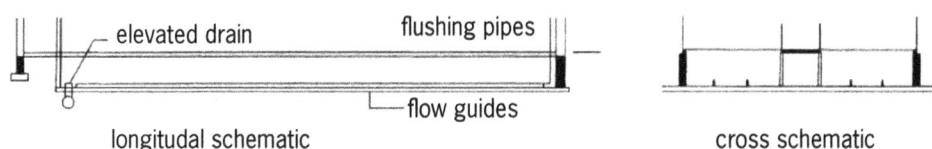

longitudal schematic   cross schematic

Emission reduction using this system

Table 1. Emission factors for flushing systems of the manure with acidified liquid manure (kg NH₃/y per pig place).

Weaners	Farrowing sows, including piglets	Dry and pregnant sows	Fattening pigs
0.16 – 0.22	3.1	1.8	0.8 - 1.1

Further references

VROM, 2006. Regeling ammoniak en veehouderij. Staatscourant 207 (October).

Infomil, http://www.infomil.nl/contents/pages/22613/bb_94-06-038_v2.pdf

Appendix 8. Cooling the manure with floating elements

System description
Ammonia emission in this system is reduced by cooling the upper layer of the manure with floating cooling elements. The heat extracted from the manure could be upgraded with a heat pump to be used for heating purposes elsewhere. The maximum temperature of the exhaust cooling water is 14 °C. The maximum temperature of the upper layer of the manure is 15 °C.

Key parameters
In order for this system to work properly, it is essential that the manure temperature should not exceed 15 °C. The cooling water is cooled down in either a soil heat exchange system or by means of a heat pump. It is important that this system is working continuously, even when the transferred heat cannot be used directly. In the latter case, surplus heat should be stored within the soil.

Key parameters during emission measurements for fattening pigs
The measuring location should (preferably) fulfil the following demands:
- The animal house should be in use for at least one fattening period.
- The animals should be kept according to the most recent legislation on animal welfare.
- The number of pigs in the pen should be between 10 and 40.
- The CO2 concentration inside the room fall below the maximum of 3,000 ppm.
- The protein content of the diet is minimal 155 g/EW for starting feed and minimal 150 g/EW for finishing feed.
- Water should be available ad libitum.
- The minimal growth of fatteners from 25-115 kg should be 750 g/d.
- The number of culled pigs is maximal 5%.
- A minimum number of 50 pigs in the room.

The following data should be registered by the farmer during the measurements:
- The total feed given per room and fattening period.
- The number of animals inside the room during the whole fattening period.
- The dates of manure removal from the manure pit.
- A list of veterinary treatments at group level.

As a check the CO_2-concentration in each room, on each farm, and in each measuring period is measured with Kitagawa tubes.

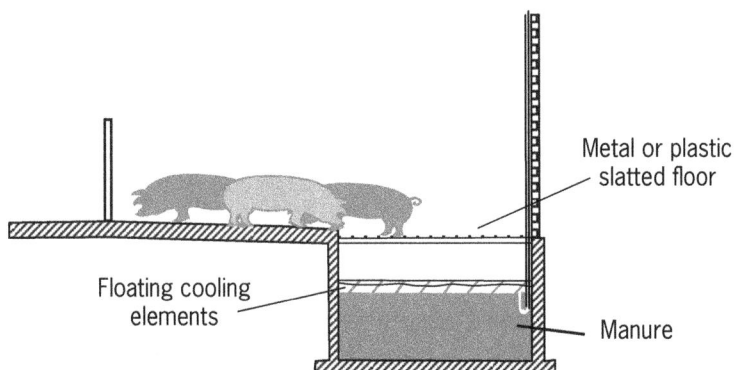

Emission reduction using this system

Table 1. Emission factors for systems with diving cooling elements in the manure pit (kg NH$_3$/y per pig place).

Weaners	Farrowing sows, including piglets	Dry and pregnant sows	Fattening pigs
0.15	2.4	2.2	1.0 - 1.4

Further references

VROM, 2006. Regeling ammoniak en veehouderij. Staatscourant 207 (October).
Infomil. http://www.infomil.nl/contents/pages/22613/bb_00-06-093.pdf

Appendix 9. Housing systems for pigs with reduced emitting surface area

System description
The main working principle of this ammonia reducing system is the reduction of the emitting manure surface area. This can be done by: (1) increasing the solid floor area and reducing the manure pit area; (2) placing slant plates in the manure pit; (3) separating the manure pit in compartments with a high manure load and with a low manure load and putting water in the compartment with a low manure load.

Key parameters
In this system, it is essential that, while reducing the emitting surface area of the manure, the fouling of other parts of the pig pen should not increase. This is especially critical when the slatted floor area is reduced and the solid floor area is increased.

Key parameters during emission measurements for fattening pigs
The measuring location should (preferably) fulfil the following demands:
- The animal house should be in use for at least one fattening period.
- The animals should be kept according to the most recent legislation on animal welfare.
- The number of pigs in the pen should be between 10 and 40.
- The CO2 concentration inside the room fall below the maximum of 3,000 ppm.
- The protein content of the diet is minimal 155 g/EW for starting feed and minimal 150 g/EW for finishing feed.
- Water should be available ad libitum.
- The minimal growth of fatteners from 25-115 kg should be 750 g/d.
- The number of culled pigs is maximal 5%.
- A minimum number of 50 pigs in the room.

The following data should be registered by the farmer during the measurements:
- The total feed given per room and fattening period.
- The number of animals inside the room during the whole fattening period.
- The dates of manure removal from the manure pit.
- A list of veterinary treatments at group level.

As a check the CO_2-concentration in each room, on each farm, and in each measuring period is measured with Kitagawa tubes.

Emission reduction using this system

Table 1. Emission factors for systems with a reduced emitting surface area (kg NH$_3$/y per pig place).

Weaners	Farrowing sows, including piglets	Dry and pregnant sows	Fattening pigs
0.17 – 0.34	2.9 – 5.0	1.8 - 2.4	1.0 - 1.4

Further references

VROM, 2006. Regeling ammoniak en veehouderij. Staatscourant 207 (October).

System description
Ammonia emission in this system is reduced by flushing the fresh manure from gutters placed in the manure pit using liquid manure. The fresh manure is flushed with a minimum frequency of two times a day. The manure gutters should cover the whole manure pit and be made of a corrosive free material, e.g. poly-ethylene. The walls of the gutter should have a slope of at least 60° and a depth of 20-60 cm. Flushing should be controlled automatically. The gutters should contain a maximum 5 cm of manure.

Key parameters
The main critical success factor of this system is the frequency of flushing.

Key parameters during emission measurements for fattening pigs
The measuring location should (preferably) fulfil the following demands:
- The animal house should be in use for at least one fattening period.
- The animals should be kept according to the most recent legislation on animal welfare.
- The number of pigs in the pen should be between 10 and 40.
- The CO_2 concentration inside the room fall below the maximum of 3,000 ppm.
- The protein content of the diet is minimal 155 g/EW for starting feed and minimal 150 g/EW for finishing feed.
- Water should be available ad libitum.
- The minimal growth of fatteners from 25-115 kg should be 750 g/d.
- The number of culled pigs is maximal 5%.
- A minimum number of 50 pigs in the room.

The following data should be registered by the farmer during the measurements:
- The total feed given per room and fattening period.
- The number of animals inside the room during the whole fattening period.
- The dates of manure removal from the manure pit.
- A list of veterinary treatments at group level.

As a check the CO_2-concentration in each room, on each farm, and in each measuring period is measured with Kitagawa tubes.

Metal slatted floor

Flushing gutters

Slurry

Emission reduction using this system

Table 1. Emission factors for systems with flushing gutters (kg NH$_3$/y per pig place).

Weaners	Farrowing sows, including piglets	Dry and pregnant sows	Fattening pigs
0.21	3.3	2.5	1.0 - 1.2

Further references

VROM, 2006. Regeling ammoniak en veehouderij. Staatscourant 207 (October).

Infomil, http://www.infomil.nl/contents/pages/22613/bb_94-06-021_v1-a_97-01-049_v1.pdf

Appendix 11. Vertical tiered cages with manure belts with forced air drying

System description

There are several designs of the battery system: flat deck, stair-step, compact- and belt-battery. The latter has cages mostly made of steel wire. They are equipped with installations for automatic watering and feeding. The bottom of the cage has an inclination, thus allowing the eggs to roll to the front. With a transport belt they are removed for further selection and packaging.

Underneath the cages runs a belt which is operated at least twice a week to remove the manure out of the house into a closed storage area. The dry matter of fresh laying hen droppings is about 22%. By blowing air over the manure, it can be dried (45-55% DM) in order to retain the ammonia inside. This is also called enriched cage system.

Cages with the dimensions of 500 mm x 450 mm x 450 mm deep can house 3 to 6 birds. With up to 8 levels (or tiers) of cages, rows exceeding 50 meters and 6 or more rows in a house, 100,000 birds can be kept in a single house. With this number of birds it is common that fully automated mechanical ventilation is used.

Key parameters

Frequent manure removal and forced air-drying are important to reduce ammonia emissions. A more frequent manure removal results in less ammonia emissions. Ammonia is generated from the breakdown of urea inside the manure. This process starts slowly, which means that by removing the manure from the house a large part of the ammonia emitting potential is removed.

The ammonia emission is also reduced when air is blown over the manure. It allows the manure to dry quickly, which stops the breakdown of urea by taking away an essential reagent: water.

In the system with the lowest ammonia emission (0.012/0.006 kg NH_3/animal place/year) the air capacity is 0.7 m^3/animal/hour (0.4 m^3 for pullets) with a temperature of 17 °C. The manure is removed every five days and has a minimum dry matter content of 55%.

Operational control for safeguarding emission reductions

Important for safeguarding emission reductions are the air capacity blown over the manure, the temperature of the air and the time period between manure removals from the house. Recording the run time of the manure belts can be used to monitor the manure removal from the house.

If either the capacity of the air or the temperature of the air are insufficient, the dry matter content of the manure will be lower than the required 55%.

Key parameters during emission measurements

Measuring periods (1996):
- first, 03 June - 30 August.
- second, 30 September - 23 December.

Feed:
- feed intake: 103.8 gr/animal/day.
- water-feed ratio: 1.8.
- OE: 2,850 - 2,880 kcal/kg.
- crude protein: 170 - 173 g/kg.

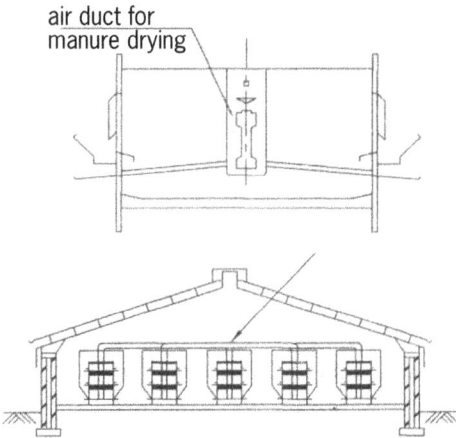

Table 1. Climate.

Period	Room temp. (°C)	Room RH (%)	Outside temp. (°C)	Outside RH (%)
1	23.9	79	17.1	61
2	22.7	76	6.9	74

Emission (reduction) using this system

Laying hens: 0.012 kg NH_3/animal place/year. Pullets: 0.006 kg NH_3/animal place/year. The emission factor for laying hens was originally established at 0.010 kg NH_3/animal place/year. Because of increasing the living space per hen by 20% the emission also was increased with 20% to 0.012 kg NH_3/animal place/year. The emission factor of pullets is derived from the factor for laying hens.

Extrapolation to similar systems and animal categories

*   Vertical tiered cages with manure belts with forced air drying (laying hens: 0.5 m³/animal/hour; 0.042 kg NH_3/animal place/year, pullets: 0.2 m³/animal/hour; 0.020 kg NH_3/animal place/year).
*   Vertical tiered cages with manure belts with whisked-forced air drying (laying hens: 0.042 kg NH_3/animal place/year, pullets: 0.020 kg NH_3/animal place/year).
*   Enriched cages (0.030 kg NH_3/animal place/year).

Further references

Reuvekamp, B.F.J. and Th.G.C.M. van Niekerk, 1997. Ammoniakemissie bij leghennen op batterijen bij drogen tot minimaal 55% drogestof en bij natte mest (Ammonia emission from manure belt batteries for laying hens with forced drying to minimal 55% dry matter and with wet manure (with English summary)). Praktijkonderzoek Pluimveehouderij, Beekbergen, pp.63.

Appendix 12. Deep litter system drying with slatted floor in manure pit

System description
The laying hens are kept in a house on the floor. In accordance to the legislation of keeping laying hens at least 1/3 of the floor area is covered with bedding and at the most 2/3 with a slatted floor. The slats are mainly made of wood or plastic (wired slats are forbidden) and are placed about 30-60 cm above the litter area. In the middle of the slatted floor the laying nests are placed. Both litter area end slatted floor are part of the usable area in the house, the surface of the laying nest is not. Per m^2 usable area 9 hens may be kept. Under the slatted floor the manure is stored for the total laying period of 13-15 months.

Air is blown through the manure in order to dry it. The air is supplied to the manure pit through the duct underneath the laying nests. From there it is blown under slats that are placed 0.10 meter above the floor of the pit. Another possible way of drying the manure is to draw air from the house through ducts passing the slatted floor where the hens are kept. Of the total surface of the slats in the manure pit 20% must be open. The maximum installed air capacity must be equal to 4.5 m^3/hen/hour. Ventilation through the manure starts at 0 m^3/hen/hour when the hens are newly placed in the house and increases up tot the maximum value at the end of the laying period.

Key parameters
For this system the key parameter is the amount of air blown through the manure. The dried manure will reach a dry matter content of about 80%. Even without forced drying, the manure will reach a dry matter content of about 80%. This is caused by spontaneous fermenting of the manure as a result of bacterial activity. This process generates heat and ammonia and can be prevented by drying the manure, thus reducing bacteria activity and urea breakdown.

Operational control for safeguarding emission reductions
Important for safeguarding emission reductions is the amount of air blown through the manure. Anemometers built into the air supply ducts are used for measuring the air capacity. These devices can also register if the forced air-drying is in operation or not.

Key parameters during emission measurements
Measuring periods (1997 - 1998):
*   first, 01 November - 31 December 1997.
*   second, 01 July - 30 August 1998.

Feed:
*   feed intake: 121.5 - 124.4 g/animal/day.
*   water-feed ratio: no information available.
*   OE / crude protein: no information available, standard commercial feed.

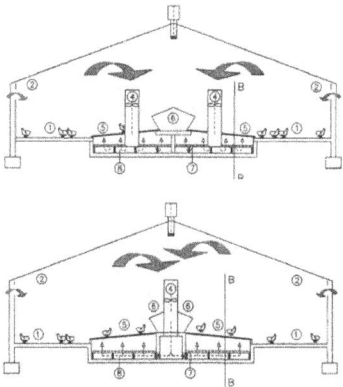

Table 1. Climate.

Period	Room temp. (°C)	Outside temp. (°C)
1	19.6	5.7
2	25.6	17.9

Emission reduction using this system

Laying hens: 0.125 kg NH_3/animal place/year. Compared with traditional housing with a emission factor of 0.315 kg NH_3/animal place/year this system gives a reduction of 60%. Pullets: no emission factor is established.

Extrapolation to similar systems and animal categories

The same system is used for broiler breeders. Only then the amount of air is 7 m^3/animal/hour with a minimum temperature of 24 °C. This system has an emission factor of 0.230 kg NH_3/animal place/year. Compared to traditional housing with an emission factor of 0.580 kg NH_3/animal place/year this system gives a reduction of 60%.

Further references

Reuvekamp, B.F.J. and Th.G.C.M van Niekerk, 1999. Mestbeluchting met buizen onder de beun (Manure drying with tubes beneath the slatted floor in a deep litter system for laying hens (with English summary)). Praktijkonderzoek Pluimveehouderij, Beekbergen, pp. 81.

Van der Haar, J.W. and R. Meijerhof, 1996. Ammoniakemissie bij vleeskuikenouderdieren in een stal met 70% roostervloer en schijnvloer in de mestput (Ammonia emission at broiler breeders in floor keeping with 70% slatted floor and extra slatted floor in the manure pit). Praktijkonderzoek Pluimveehouderij, Beekbergen, pp. 51.

Appendix 13. Deep litter system with manure drying from above

System description

The laying hens are kept on the floor of a house. In accordance to the legislation regarding the keeping of laying hens, at least 1/3 of the floor area is covered with bedding and at the most 2/3 of the floor is slatted. The slats are mainly made of wood or plastic (wired slats are forbidden) and are placed 30-60 cm above the litter area. In the middle of the slatted floor the laying nests are placed. Both litter and slatted floor area are part of the usable area in the house, the surface area of the laying nest is not. Per m^2 usable area 9 hens may be kept.

Under the slatted floor the manure is stored for the total laying period of 13-15 months. Similar to battery houses, the manure is dried by blowing air over it through tubes. Per hen 1.2 m^3/hour of 20 °C air is used. This air is heated either in an air mixer or a heat exchanger.

Key parameters

For this system the important key parameters are the amount of drying air and its temperature. The dried manure will reach a dry matter content of about 80%. Even without forced drying, the manure will reach a dry matter content of about 80%. This is caused by spontaneous fermenting of the manure as a result of bacterial activity. This process generates heat and ammonia and can be prevented by drying the manure, thus reducing bacteria activity and urea breakdown.

Operational control for safeguarding emission reductions

Important parameters for safeguarding emission reductions are the amount of drying air blown over the manure and its temperature. Anemometers built into the air supply ducts are used for measuring the air capacity. These devices can also register if the forced air-drying is in operation or not. The temperature of the air can be measured using a thermometer.

Key parameters during emission measurements

Measuring periods (1997):
* first, 01 June - 31 August.
* second, 01 October - 31 December.

Feed:
* feed intake: 122.1 gr/animal/day.
* water-feed ratio: 1.79.
* OE: 2,740 kcal/kg.
* crude protein: 155 g/kg.

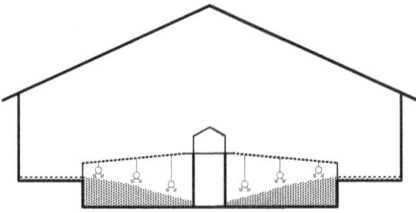

Table 1. Climate.

Period	Room temp. (°C)	Room RH (%)	Outside temp. (°C)	Outside RH (%)
1	23.9	57	18.1	82
2	20.2	55	7.0	93

Emission reduction using this system

Laying hens: 0.125 kg NH_3/animal place/year. Compared to traditional housing with an emission factor of 0.315 kg NH_3/animal place/year, this system gives an ammonia emission reduction of 60%. Pullets: no emission factor is established.

Extrapolation to similar systems and animal categories

The same system is used for broiler breeders. There, the amount of drying air is 2.5 m^3/animal/ hour with a minimum temperature of 24 °C. The emission factor for this system is 0.250 kg NH_3/ animal place/year (a 57% reduction compared to traditional housing with 0.580 kg NH_3/animal place/year).

Further references

Reuvekamp, B.F.J. and Th.G.C.M van Niekerk, 1999. Mestbeluchting met buizen onder de beun (Manure drying with tubes beneath the slatted floor in a deep litter system for laying hens (with English summary)). Praktijkonderzoek Pluimveehouderij, Beekbergen, pp. 81.

Van der Haar, J.W., R Meijerhof and J.H. Middelkoop, 1998. Ammoniakemissie bij vleeskuikenouderdieren in grondhuisvesting met mestbeluchting van bovenaf (Ammonia emission at Broiler Breeders in floor keeping with forced air drying (with English summary)). Praktijkonderzoek Pluimveehouderij, Beekbergen, pp. 70.

Appendix 14. Aviary system (perchery)

System description

The birds are kept in a house with litter on the floor and scaffolds with several levels. The levels are made of slatted floor with a manure belt underneath. Also perches are present in the top of the scaffolds. The slats are mainly made of wood or plastic (wired slats are forbidden). Divided over the house laying nest are placed in or between the scaffolds. In accordance to the legislation of keeping laying hens, at least 1/3 of the concrete floor area is covered with bedding. Both litter area and slatted floors are part of the usable area in the house, the surface of the laying nests is not. Per m² usable area, a maximum of 9 hens may be kept.

Similar to battery systems, the manure on the belts can be dried by blowing air over it. The air is blown over the manure using tubes. The capacity and the temperature of the air are different. The air is heated either in an air mixer or a heat exchanger.

The manure on the belts is removed from the house at least once a week. The litter on the floor stays in the house for the whole laying period of 13-15 months.

Key parameters

For this system the key parameters are the percentage of slatted floor, the amount of air blowing over the manure and its temperature, and the frequency of manure removal using the belts. Each combination of these parameters results in a different emission.

Operational control for safeguarding emission reductions

Important for safeguarding emission reductions are the percentage slatted floor, the air capacity blown over the manure and the temperature of the air. Anemometers built into the air supply ducts are used for measuring the air capacity. These devices can also register if the forced air-drying is in operation or not. The temperature of the air can be measured using a thermometer. The percentage of slatted flooring can be calculated form the technical drawings necessary for getting a permit.

Key parameters during emission measurements

The literature describing the measurements on the first aviary systems (1991) did not hold information on animal feed and climate. Only information on the key parameters from the newest systems is available (see References). An example of data measured is given in Table 1.

Emission reduction using this system

The first measurements (1991) resulted in an emission factor of 0.090 kg NH_3/animal place/year. The other three have an emission factor of 0.025, 0.037 and 0.055 kg NH_3/animal place/year. This gives a reduction of respectively 72, 59 and 39%.

For pullets the emission factors are established at 0.014, 0.020 and 0.030 kg NH_3/animal place/year for the systems with the same equipment as used with laying hens. The derived emission factor for the first made system was calculated at 0.050 kg NH_3/animal place/year. The reduction is the same as with aviaries for laying hens.

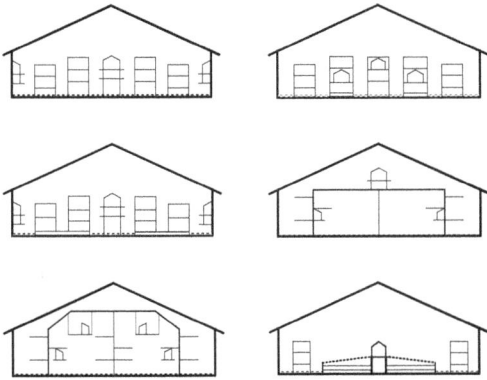

Table 1. Key parameters from the newest systems.

Measuring period

	Start	End	T house (°C)	T outside (°C)	RH house (%)	RH outside (%)	DM manure (%)
1	30-6-2001	29-8-2001	24	18.3	63.5	78	43
2	1-10-2001	30-11-2001	20.9	10	64.9	88.1	39.9

Feed	Intake (g/a/d)	118.0	Amount of air over the manure belts (m3/a/h)	0.2
	W/f-ratio	1.61	Average temperature of drying air (°C)	19.2
	OE (kcal/kg)	2850	Frequency of manure removal (times/week)	2
	RE (g/kg)	161.1-153.8		

Further references

Beurskens, A.G.C., J.M.G. Hol and G. Mol, 2002. Onderzoek naar de ammoniak- en geuremissie van stallen LV, Volièrestal voor leghennen. (Aviary housing system for laying hens (with English summary)) Rapport 2002-16, IMAG, Wageningen.

Blokhuis, H.J. and J.H. Metz, 1994. Volièrehuisvesting voor leghennen. Spelderholt Uitgave No. 627, ID-DLO, Beekbergen / IMAG-DLO Rapport No. 95-5, IMAG, Wageningen.

Van Emous, R.A., B.F.J. Reuvekamp and Th.G.C.M. Fiks-van Niekerk, 2001. Verlichtings-, ammoniak-, stof-, en arbeidsonderzoek bij twee volièresystemen (Lighting, ammonia, dust and labour research of two aviary housing systems (with English summary)). Rapport 235, Praktijkonderzoek Veehouderij, Lelystad.

Appendix 15. Deep litter system in a house with two or more stories

System description
In this system the house consist of two or more floors. In each floor the laying hens are kept. In accordance to the legislation of keeping laying hens, at least 1/3 of the floor area is covered with bedding and at the most 2/3 of the floor is slatted. The slats are mainly made of wood or plastic (wired slats are forbidden) and is placed 50 cm above the litter area. In the middle of the slatted floor the laying nests are placed. Both litter area and slatted floor are part of the usable area in the house, the surface of the laying nest is not. Per m² usable area, 9 hens may be kept. Under the slatted floor manure belts are installed to remove the manure out of the house at a frequency of twice a week. There is no forced drying of the manure.

Key parameters
For this system the key parameter is frequency at which the manure is removed from the house. The more often the manure is removed from the house, the lower the ammonia emission will be. Ammonia is generated from the urea in the manure. This process starts slowly, which means that by removing the manure from the house a large part of the ammonia emitting potential is removed.

Operational control for safeguarding emission reductions
Important for safeguarding emission reductions is the time period between manure removals from the house. Recording the run time of the manure belts can be used to monitor the manure removal from the house.

Key parameters during emission measurements
Measuring periods (2000):
- first, 01 July - 31 August.
- second, 01 October - 30 November.

Feed:
- feed intake: 109 - 113 gr/animal/day.
- water-feed ratio: 1.63 - 1.70.
- OE: 2,842 kcal/kg.
- crude protein: 175 g/kg.

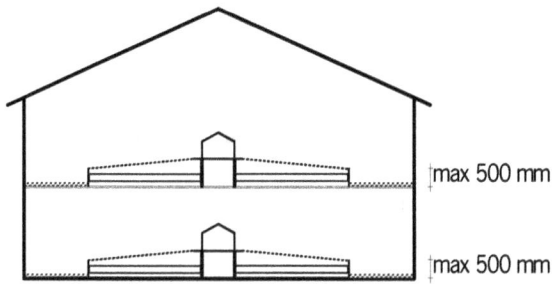

max 500 mm

max 500 mm

Table 1. Climate.

Period	Room temp. (°C)		Room RH (%)		Outside temp. (°C)	Outside RH (%)
	First floor	Second floor	First floor	Second floor		
1	23.0	24.7	64	60	19.2	75
2	21.9	21.5	59	61	10.8	89

Emission reduction using this system

Laying hens: 0.068 kg NH₃/animal place/year. Compared to traditional housing with an emission factor of 0.315 kg NH₃/animal place/year this system gives an ammonia emission reduction of 78%. Pullets: no emission factor is established.

Further references

Hol, J.M.G., P. de Gijsel and P.W.G. Groot Koerkamp, 2001. De ammoniak- en geuremissie van een scharrelstal met twee verdiepingen met mestbanden onder de roosters (The ammonia and odour emission from a deep litter house with two stories and manure belts under the slatted floor (with abstract in English)). Rapport 01-01, IMAG, Wageningen.

Appendix 16. Deep litter system with manure drying with vertical tubes

System description
The broiler breeders are kept on the floor in a house. In accordance to the legislation of the Product Board for Poultry and Eggs (PPE) at least 1/3 of the floor area is covered with bedding and at the most 2/3 of the floor area is slatted. The slats are mainly made of wood or plastic (wired slats are forbidden) and placed about 30-60 cm above the litter area. In the middle of the slatted floor the laying nests are placed. The total surface of the house is the usable area; the surface of the laying nest is included. Per m² usable area about 7.5 birds (hens and cocks) may be kept. Under the slatted floor the manure is stored for the total laying period of 10-11 months.

The manure is dried by blowing air through it. Therefore tubes with holes in it are placed in the manure. The tubes are placed according to a specific plan and also the holes in the tubes have specific dimensions and placing. The air is supplied through ducts or pipes from underneath the laying nests. The channel under the nests is used to extract the warm air. Per hen 1 m³ air/hour is used. The temperature of the air is not allowed to get higher than the temperature of the manure, to avoid condensation of water inside the manure.

Key parameters
For this system the important key parameters are the amount of air and its temperature. The dried manure will reach a dry matter content of about 80%. Even without forced drying, the manure will reach a dry matter content of about 80%. This is caused by spontaneous fermenting of the manure as a result of bacterial activity. This process generates heat and ammonia and can be prevented by drying the manure, thus reducing bacteria activity and urea breakdown.

Operational control for safeguarding emission reductions
Important for safeguarding emission reductions are the amount of air blown through the manure and its temperature. Anemometers built into the air supply ducts are used for measuring the air capacity. These devices can also register if the forced air-drying is in operation or not. The temperature of the air can be measured using a thermometer.

Key parameters during emission measurements
Measuring periods (2000/2001):
- first, 14 October 2000 - 11 December 2000.
- Second, 02 June 2001 - 31 August 2001.

Feed:
- feed intake: 44.7 kg per placed hen.
- water-feed ratio: 1.8.
- crude protein: 156-165 g/kg.

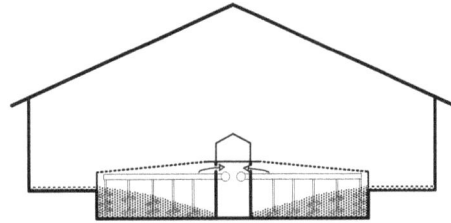

Table 1. Climate.

Period	Room temp. (°C)	Room RH (%)	Outside temp. (°C)	Outside RH (%)
1	20.6	66	9.7	86
2	23.4	65	19.5	73

Emission reduction using this system

Broiler breeders: 0.435 kg NH_3/animal place/year. (a 25% reduction compared to traditional housing with 0.580 kg NH_3/animal place/year). Pullets: no emission factor is established

Extrapolation to similar systems and animal categories

Instead of blowing air through the manure, other systems blow the air over the manure. Only more air is used with a higher temperature.

Further references

Scheer, A., J.M.G. Hol and G. Mol, 2002. Stal voor vleeskuikenouderdieren met continue drogen van mest (Housing system for broiler breeders with continuous drying of manure (with English summary)). IMAG-Rapport 2002-15, Wageningen.

Appendix 17. Floor heating and cooling (Kombidek®)

System description

The housing is very similar to traditional housing. The broilers (18- 24 per m²) are kept in a house on a concrete floor with bedding material. The house and the floor are insulated and mechanically ventilation is used. The floor contains heat exchange elements. Through these elements warm or cold water is running. Warm water is used in the first week of the growing period, when the birds need higher temperatures. From day 21 onward, cold water is used to cool down the floor and the litter on top of it. In this way bacterial activity is decreased and ammonia production is reduced as compared to that in a traditional housing system. The warm water coming from the floor is running through a heat pump and the heat coming out of the heat pump stored in the soil beneath or next to the house. After the growing period of about 42 days the manure is removed from the house and the house is cleaned. The heat from the underground storage is used to warm the floor for the next flock.

Key parameters

For this system the key parameter is the temperature of the floor (or the litter). Without the cooling in the second half of the growing period the manure starts to heat up by it self. This is caused by spontaneous fermenting of the manure as a result of bacterial activity. This process generates heat and ammonia and can be prevented by cooling the manure, thus effectively reducing bacteria activity and urea breakdown.

Operational control for safeguarding emission reductions

Important for safeguarding emission reductions is the temperature of the floor. The temperature can be measured with a thermometer.

Key parameters during emission measurements

Table 1. Diets during measurements.

Period	Data measuring period	Feed intake (kg/bird)	Water-feed ratio	OE (MJ/kg)	Crude protein (%)
1	07/10 - 08/25 1997	3.73	1.85	11.5-12.6	22.4-20.1
2	09/01 - 10/14 1997	3.51	1.76	11.5-12.6	22.4-20.1
3	10/24 - 12/10 1997	3.75	1.76	11.5-12.6	22.4-20.1
4	07/23 - 09/02 1998	3.33	1.79	11.5-12.4	22.7-19.7

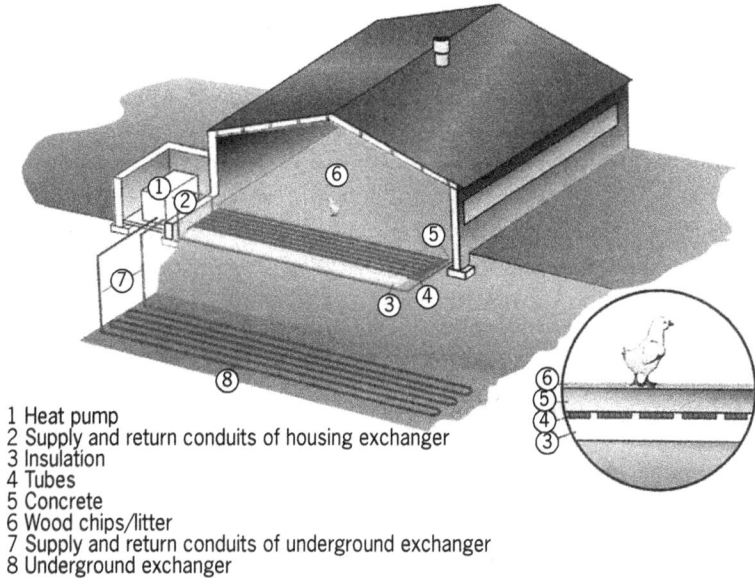

1 Heat pump
2 Supply and return conduits of housing exchanger
3 Insulation
4 Tubes
5 Concrete
6 Wood chips/litter
7 Supply and return conduits of underground exchanger
8 Underground exchanger

Table 2. Climate/litter.

Period	Room temp. (°C)	Room RH (%)	Outside temp. (°C)	Outside RH (%)	Dry matter content litter (%)	
					< 21 days	> 21 days
1	26.8	61	19.6	76	60.3	49.9
2	24.9	60	13.8	78	76.8	51.8
3	24.6	58	4.8	83	71.6	50.5
4	25.8	59	17.1	78	76.4	63.4

Emission reduction using this system

The emission factor is 0.045 kg NH_3/animal place/year. Compared to traditional housing with 0.080 kg NH_3/animal place/year this is an ammonia emission reduction of 44%.

Further references

Hol, J.M.G. and P.W.G. Groot Koerkamp, 1998. Praktijkonderzoek naar de ammoniakemissie van stallen XXXX: Vleeskuikenstal met verwarming en koeling van de vloer (Broiler house with heating and cooling of the floor (with English summary)). Rapport 98-1004, IMAG, Wageningen.

Appendix 18. Mixed air ventilation

System description

The mixed air ventilation housing is very similar to traditional housing. The broilers (18- 24 per m²) are kept in a house on a concrete floor with bedding material. The house is insulated and ventilated mechanically. The house contains ventilation shafts to circulate the warm air from the ceiling down to the ground. Per 150 m² there is one shaft with a capacity of 1,8 m³/hour/bird at 0 Pa overpressure. The downward circulation starts off at 0 m³ on the first day of the growing period and increases linear to maximum capacity on day 42, the end of the growing period. Afterwards, the broilers and the manure are removed from the house and the house is cleaned.

Key parameters

For this system the important key parameter is the downward circulation capacity.

Operational control for safeguarding emission reductions

Important for safeguarding emission reductions is the capacity of the circulation. These parameters are stored in the equipment that regulates the fans.

Key parameters during emission measurements

Table 1. Diets during measurements.

Period	Data measuring period	Live weight (g)	Feed conversion (kg/kg at 1,500 g)	Water-feed ratio	OE (MJ/kg)	Crude protein (%)
1	06/24 - 07/24 2002	2,328	1.28	1.73	11.9-12.7	22.1-19.5
2	10/02 - 11/02 2002	2,107	1.33	1.65	11.9-12.7	22.1-19.5
3	03/22 - 05/03 2005	2,262	1.36	1.78	–	22.1 - 19.1
4	06/30 - 11/02 2005	2,340	1.47	1.89	–	22.1 - 19.1

Table 2. Climate/litter.

Period	Room temp. (°C)	Room RH (%)	Outside temp. (°C)	Outside RH (%)	dm-content litter (%)
1	26.8	69	21.6	77	58.6
2	24.9	65	11.6	89	61.7
3	24.7	58	13.2	75.5	
4	25.7	85	19.7	88	

Emission reduction using this system

Emission factor is 0.037 kg NH_3/animal place/year. Comparing with traditional housing with 0.080 kg NH_3/animal place/year this is a reduction of 54%.

Further references

Bleeker, A. and W.C.M. van den Bulk, 2005a. Verificatie ImagO stalsysteem. ECN-C--05-053, ECN, Petten.

Bleeker, A. and W.C.M. van den Bulk, 2005b. Tweede verificatie ImagO stalsysteem. ECN-C--05-079, ECN, Petten.

Scheer, A., J.M.G. Hol and G. Mol. 2003. Stal voor vleeskuikens met vloerverwarming en mixluchtventilatoren voor het drogen van de strooisellaag (House for broilers with floor heating and air-mixing for drying litter (with English summary)). IMAG Rapport 2003-15, Wageningen.

System description

In this system, a multi-tier housing cage system for broilers is used. In each tier, the birds stay on top of a conveyor belt that is covered with paper and litter. The birds stay inside the tier for the entire growing period. At the end of the growing period, the manure belt is activated and the birds are transported to the end of the house, where birds and manure are automatically separated when leaving the house.

The ventilation air is pulled through the tiers, causing the litter to dry. The incoming air on the side of the tier is heated or cooled at a central point when entering the house. Reduction of the ammonia emission is caused by drying the litter and keeping the temperature of the litter low.

Key parameters

For this system the important key parameter is the capacity of the ventilation.

Operational control for safeguarding emission reductions

Important for safeguarding emission reductions is the capacity of the ventilation. These parameters are stored in the equipment that regulates the fans.

Key parameters during emission measurements

Measuring periods (2004):
• first, 02 July - 09 August.
• second, 02 October - 7 November.

Feed:
• feed intake: 3.18 / 3.98 kg/bird.
• water-feed ratio: 1.92 / 1.80.
• OE: 11.9 - 12.6 MJ/kg.
• crude protein: 20.7 - 22.5 g/kg.

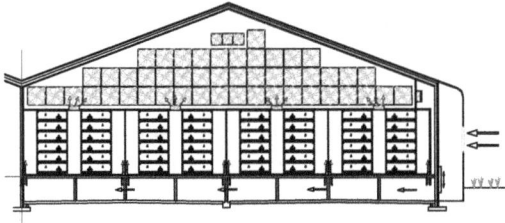

Table 1. Climate.

Period	Room temp. (°C)	Room RH (%)	Outside temp. (°C)	Outside RH (%)
1	28.4	62	21.2	72
2	29.6	55	13.2	83

Emission reduction using this system

Emission factor is 0.020 kg NH_3/animal place/year. Comparing with traditional housing with 0.080 kg NH_3/animal place/year this is a reduction of 75%.

Further references

Huis in 't Veld, J.W.H., S.G. Van der Top, J.M.G. Hol and J. Mosquera, 2005. Onderzoek naar de ammoniak- en geuremissie van stallen LXIII: Meeretagesysteem voor vleeskuikens (Multi tier house for broilers (with English summary)). Rapport 367, Agrotechnology & Food Innovations B.V., Wageningen.

System description

In a biological air scrubber, ventilation air is passed either in counter or crossflow over a packed bed that is continuously wetted by spraying a recirculating water phase containing an indigenous population of bacteria species. The bed provides the necessary air to fluid surface area for the mass transfer of oxygen, ammonia and odour compounds to the watery phase. The bed is often made of inert plastic and comprises of either a loose network or ribboned channels, both with a low specific volume (< 5%) to ensure a low pressure drop due to drag.

Breakdown of the contaminants largely takes place in the collection basin under the packed bed. Ammonia is digested by nitrification bacteria and converted into nitrite (NO_2^-) and nitrate (NO_3^-). A high concentration of intermediate and end products can lead to poisoning of the biomass, resulting in poor emission abatement capabilities. The maximum concentration of these compounds (together with excess biomass) is regulated by purging part of the watery phase.

As an alternative to purging, denitrification can be achieved in a separate storage tank under anoxic conditions. Denitrification bacteria are capable of converting nitrite and nitrate to N_2, provided that these bacteria are fed with a suitable carbon source that is consumed during this process. Purging is thus minimised to only decrease the excess biomass formed.

Key parameters

The primary concern of operating a biological air scrubber are the chemical conditions of the watery phase containing the biomass. In a well functioning biological air scrubber, the concentration ratio of captured ammonia and the sum of concentrations nitrate and nitrite is within the range [0.8; 1.2]. This guarantees a sufficient bacteriological acid production to protonate the ammonia.

$$M_{N/N} = \frac{[NH_4^+]}{[NO_3^-] + [NO_2^-]} \approx 1$$

Typical operating pH is about 7. Generally it should fall between 6.5 and 7.5. The maximum ammonium concentration allowed in the system is about 3.2 g N/l, whereas the total N value in the watery phase should exceed 1 g/l.

Operational control for safeguarding emission reductions

Main operation control parameter is the $M_{N/N}$ ratio, which is periodically determined in a certified chemical laboratory. Values of $M_{N/N}$ lower than 0.8 indicate a sudden (unusual) drop in ammonia load just prior to sampling. At values bigger than 1.2, the desired nitrification is suffering, yet with a normal pH value, enough ammonia is still removed. Values above 5 indicate that nitrification has nearly stopped and that the system is failing.

By demanding a minimum nitrogen concentration of 0.8 g N/l it is certain that acquired ammonia reductions are reached by a sufficient biological activity. The purge flow is adjusted to accommodate for this figure. The flow settings are recorded for control purposes.

Additionally, the following operational control should be guaranteed:

- Samples of the watery phase should be taken twice per year by qualified personnel for analysis in a certified laboratory.
- The packed bed should be cleaned once per year to prevent an excessive pressure drop and subsequent high energy costs.
- A log should be kept stating all measurements, their results and all maintenance issues.
- Continuous and tamper free registration of the purge flow pH and the active time of the recirculation pump for a mandatory weekly check of the system.
- A maintenance contract with the supplier of the system should cover yearly maintenance and day to day advice regarding normal system operations.

Key parameters during emission measurements

The biological air filter is used as an end-of-pipe technique and can be used on a number of animal husbandry systems for cleaning exhaust air. During measurements, it is essential that the ammonia load is constant and that the system is allowed to reach its dynamic equilibrium for at least two weeks. Only in this way, a closed N-mass balance can be produced from the measurements.

Emission reduction using this system

Biological air scrubbers are capable of reaching 70% ammonia reduction.

Extrapolation to similar systems and animal categories

This system is applicable for all animal categories. When the treated air contains a lot of dust (i.e. exhaust air from hen houses), special precautions should be taken to prevent clogging of the system.

Further references

Scholtens, R. and R.W. Melse, 2004. Inspectie van luchtwassystemen voor mechanisch geventileerde varkens- en pluimveestallen, IMAG report (in Dutch).

Appendix 21. Chemical air scrubber

System description
In a chemical air scrubber, all ventilation air is passed either in counter or crossflow over a packed bed that is continuously wetted by spraying a recirculating acidified water phase. The bed provides the necessary air to fluid surface area for the mass transfer of ammonia to the watery phase. The bed is often made of inert plastic and comprises of either a loose network or ribboned channels, both with a low specific volume (< 5%) or lamellae, to ensure a low pressure drop due to drag.

The watery phase is acidified to pH values in the range of [1-4.5] using concentrated sulfuric acid. Prior to passing the packed bed, dust is removed from the air flow by wet washing using the same acidified water phase.

Key parameters
The primary concerns of operating a chemical air scrubber are the recirculating flow across the packed bed, the pH of the watery phase and the purge flow.

Operational control for safeguarding emission reductions
- Samples of the watery phase should be taken twice per year by qualified personnel for analysis in a certified laboratory.
- The packed bed should be cleaned once per year to prevent an excessive pressure drop and subsequent high energy costs.
- A log should be kept stating all measurements, their results and all maintenance issues.
- Continuous and tamper free registration of the purge flow pH and the active time of the recirculation pump for a mandatory weekly check of the system.
- A maintenance contract with the supplier of the system should cover yearly maintenance and day to day advice regarding normal system operations.

Key parameters during emission measurements

The chemical air scrubber is used as an end-of-pipe technique and can be used on a number of animal husbandry systems for cleaning exhaust air. During measurements, it is essential that the ammonia load is constant and that the system is allowed to reach its dynamic equilibrium for at least two weeks. Only in this way, a closed N-mass balance can be produced from the measurements.

Emission reduction using this system

Chemical air scrubbers are capable of reaching 95% ammonia reduction.

Extrapolation to similar systems and animal categories

This system is applicable for all animal categories. When the treated air contains a lot of dust (i.e. exhaust air from hen houses), special precautions should be taken to prevent clogging of the system.

Further references

Scholtens, R. and R.W. Melse, 2004. Inspectie van luchtwassystemen voor mechanisch geventileerde varkens- en pluimveestallen, IMAG Report (in Dutch).

List of authors

André J.A. Aarnink
Animal Sciences Group of Wageningen UR
P.O. Box 65, 8200 AB Lelystad, The Netherlands
e-Mail: andre.aarnink@wur.nl
Telephone: +31 (0)320-293589

Hilko H. Ellen
Animal Sciences Group of Wageningen UR
P.O. Box 65, 8200 AB Lelystad, The Netherlands
e-Mail: hilko.ellen@wur.nl
Telephone: +31 (0)320-293504

Jan F.M. Huijsmans
Plant Research International
P.O. Box 16, 6700 AA Wageningen, The Netherlands
e-Mail: jan.huijsmans@wur.nl
Telephone: +31 (0)317-476310

Gideon Kruseman
Agricultural Economics Research Institute
P.O. Box 29703, 2502 LS, The Hague, The Netherlands
e-Mail: gideon.kruseman@wur.nl
Telephone: +31 (0)70-3358189

Harry H. Luesink
Agricultural Economics Research Institute
P.O. Box 29703, 2502 LS, The Hague, The Netherlands
e-Mail: harry.luesink@wur.nl
Telephone: +31 (0)70-3358315

Julio Mosquera Losada
Animal Sciences Group of Wageningen UR
P.O. Box 65, 8200 AB Lelystad, The Netherlands
e-Mail: julio.mosquera@wur.nl
Telephone: +31 (0)320-293571

Oene Oenema
Alterra – Soil Science Centre
P.O. Box 47, 6700 AA Wageningen, The Netherlands
e-Mail: oene.oenema@wur.nl
Telephone: +31 (0)317-486483

Michel C.J. Smits
Animal Sciences Group of Wageningen UR
P.O. Box 65, 8200 AB Lelystad, The Netherlands
e-Mail: michel.smits@wur.nl
Telephone: +31 (0)320-293596

Dick A.J. Starmans
Animal Sciences Group of Wageningen UR
P.O. Box 17, 6700 AA Wageningen, The Netherlands
e-Mail: dick.starmans@wur.nl
Telephone: +31 (0)317-476312

Klaas W. Van der Hoek
National institute for public health and the environment, RIVM
P.O. Box 1, 3720 BA Bilthoven, The Netherlands
e-Mail: klaas.van.der.hoek@rivm.nl
Telephone: +31 (0)30-2743774

Index

www.ingramcontent.com/pod-product-compliance
Lightning Source LLC
Chambersburg PA
CBHW081458190326
41458CB00015B/5280